T0331138

THE EARTH'S INNER CORE
Revealed by Observational Seismology

The inner core is a planet within a planet: a hot sphere with a mass of 100 quintillion tons of iron and nickel that lies more than 5000 km beneath our feet. It plays a crucial role in driving outer core fluid motion and the geodynamo, which generates the Earth's magnetic field. This book is the first to provide a comprehensive review of past and contemporary research on the Earth's inner core from a seismological perspective. Chapters cover the collection, processing, and interpretation of seismological data, as well as our current knowledge of the structure, anisotropy, attenuation, rotational dynamics, and boundary of the inner core. Reviewing the latest research and suggesting new seismological techniques and future avenues, it is an essential resource for both seismologists and non-seismologists interested in this fascinating field of research. With electronic supplements, including inner core-sensitive datasets and 3D visualisations, it will also form a useful resource for courses in seismology and deep Earth processes.

HRVOJE TKALČIĆ is an associate professor in the Seismology and Mathematical Geophysics Group at The Australian National University. He has authored over 60 research papers, and his research interests include the structure and dynamics of the Earth's interior using observational seismology and mathematical geophysics. His recent projects focus on developing new approaches in lithospheric and mantle imaging and studying seismic sources. Dr. Tkalčić is a manager of the Warramunga Seismic and Infrasound Array in the Northern Territory, Australia, and he also participates in improving global coverage of seismic data by deployment in remote regions.

THE EARTH'S INNER CORE

Revealed by Observational Seismology

HRVOJE TKALČIĆ

The Australian National University, Canberra

CAMBRIDGE
UNIVERSITY PRESS

CAMBRIDGE
UNIVERSITY PRESS

Shaftesbury Road, Cambridge CB2 8EA, United Kingdom

One Liberty Plaza, 20th Floor, New York, NY 10006, USA

477 Williamstown Road, Port Melbourne, VIC 3207, Australia

314–321, 3rd Floor, Plot 3, Splendor Forum, Jasola District Centre, New Delhi – 110025, India

103 Penang Road, #05–06/07, Visioncrest Commercial, Singapore 238467

Cambridge University Press is part of Cambridge University Press & Assessment, a department of the University of Cambridge.

We share the University's mission to contribute to society through the pursuit of education, learning and research at the highest international levels of excellence.

www.cambridge.org
Information on this title: www.cambridge.org/9781107037304

DOI: 10.1017/9781139583954

First published 2017

A catalogue record for this publication is available from the British Library

Library of Congress Cataloging-in-Publication data
Names: Tkalčić, Hrvoje, author.
Title: The earth's inner core : revealed by observational seismology / Hrvoje Tkalčić.
Description: Cambridge, United Kingdom ; New York, NY : Cambridge University Press, 2017.
| Includes bibliographical references and index.
Identifiers: LCCN 2016042510 | ISBN 9781107037304 (hardback)
Subjects: LCSH: Earth (Planet) – Core. | Seismology.
| BISAC: SCIENCE / Geophysics.
Classification: LCC QE509.2 .T53 2017 | DDC 551.1/12–dc23
LC record available at https://lccn.loc.gov/2016042510

ISBN   978-1-107-03730-4   Hardback

Contents

Colour plate section can be found between pages 84 and 85

Preface

"Every day is a journey, and the journey itself is a home."

Matsuo Bashō, *Narrow Road to the Interior* (Bashō, 1694)

In Jules Verne's novel, *Journey to the Centre of the Earth* (Verne, 1864), a forest of giant mushrooms grows on the shores of a large subterranean sea, ichthyosaurs and plesiosaurs clash for survival in its murky waters, and the heavy strata of the atmosphere permeates the gigantic cavern that must have somehow collapsed inwards from the surface. One and a half centuries later, our imagination for the deepest of the depths does not cease to exist and 'the people of perpendiculars',[1] like Professor Lindenbrock from Verne's novel, whose only wish is to slide down the Earth's radius, are luckily still among us. Verne's phantasmagorical landscapes, however, get washed away by the scientific rigour of the deep Earth images that emerge from seismological studies. This is not to say that the Earth's centre in our contemporary view is featureless. Quite to the contrary, the inner core is a planet within a planet: a hot sphere with a mass of one hundred quintillion tons of iron and nickel that lies about 5150 kilometres beneath our feet, still waiting to be 'discovered'. Its surface seems to be rough and mushy in places, and its interior deformed by internal stresses. It might contain heterogeneity at various scales, and host another smaller shell more elusive to our seismological probes. Despite its small volume (less than 1 per cent of the Earth's volume), the Earth's inner core contains about 10 per cent of the total magnetic field energy. It plays a crucial role in outer core fluid motions and the geodynamo, which generates the Earth's magnetic field. Without the magnetic field, life on Earth would be impossible. Embedded in the liquid outer core with a strong magnetic field and exposed to large gravitational pulling from the surrounding mantle, it spins faster than the mantle, and, at times, it slows down. But how do we know all this?

---

[1] The term used by William Butcher to refer to Jules Verne himself, in an introduction to *Journey to the Centre of the Earth*, Oxford University Press, 1992.

This book attempts to answer that question as it skips a significant part of 'the journey' and arrives by means of seismic waves at our destination – the inner core – right from the start. As seismic waves produced by earthquakes, explosions, and other natural phenomena reverberate through the solid Earth, they are reflected or scattered from discontinuities within and between the crust, mantle, and inner and outer core. Changes in the composition and temperature of Earth's minerals cause the waves to change their speed, bend, and even reverse their paths, all of which is manifested in recorded seismograms. The waves that traverse the inner core are thus the only direct probe available, and they become a subject of study.

Due to the exponential growth of geophysical data in recent years, it has become possible to image the Earth's deep interior, albeit not quite to the level of detail obtained for structure near the Earth's surface. Seismological observations have been the pivotal points for major advances in our understanding of the deepest Earth structure, which more often lead than follow geodynamical predictions. Indeed, global observational seismology is a powerful tool that serves as an inverted telescope with which we can probe the deepest parts of the Earth's interior, including the inner core. However, probing the inner core with seismic waves is similar to looking at a distant object in the universe through an imperfect lens that distorts the image. Imperfect knowledge of the rest of the Earth's layers is projected to the centre as a challenge that will be overcome only by innovative ideas and methods, some of which are presented in this book. Indeed, most seismologists would confirm that we are still in a discovery rather than mapping stage in our pursuit of understanding the deep Earth. This makes studying the Earth's inner core a fascinating subject.

Despite the fact that seismology leads the observational efforts, major advances in our understanding of the Earth's inner core would have not been possible without combining seismological observations with the results from geodynamics, magnetohydrodynamic modelling, mineral physics, and mathematical geophysics. However, interactions between researchers from different disciplines are often limited to special sessions on the Earth's core at international meetings and conferences. The geophysical values derived from seismological analyses (e.g. the inner core to outer core density ratio, the strength of anisotropy in the inner core, or the amount of differential rotation with respect to the mantle) are often utilised in geodynamical considerations or modelling without sufficient scrutiny and clear understanding of how these quantities were derived. This is in part because the inner core is buried deep below our feet and is difficult to scrutinise due to a lack of experimentally controlled conditions. It is also because of the level of specialisation required and a perceived lack of data visibility. Despite the importance of the subject, the communication among various groups of researchers still remains relatively undeveloped. We live in times when science can no longer thrive when fragmented into a large number of sub-disciplines. Studies of the Earth's inner core are no exception. Hence, there is a need more than ever for specialists from various disciplines to collaborate and share seismological data in our common quest to elucidate Earth's inner core.

Understanding inner core structure and dynamics, including energy exchange across the liquid core boundaries, helps Earth and Planetary Scientists to better understand planetary formation, the workings of the Earth's magnetic field, and the age of the inner core, the time capsule to understanding Earth's past, present, and future. During the past several decades, our understanding of the inner core's internal structure and dynamics has completely changed owing to modern observational seismology and the expansion of worldwide seismographic stations. Yet, there are no contemporary monographs available that place emphasis on the Earth's inner core from a seismological perspective. The *AGU Geodynamics Series on Earth's Core, Dynamics, Structure, Rotation*, vol. 31, published 14 years ago, is a collection of review papers on the subject, but they focus on selected aspects of the Earth's core, and therefore, a connection among the presented material is difficult to establish. A number of major advances have been made since its publication. Some basic facts about the inner core are present in a number of seismological books at various levels, but the inner core has never been presented as a stand-alone subject in spite of its importance and great interest among students.

My intention has been to look at the Earth's inner core from the perspective of an observational seismologist and present the methods and seismological data that have been used to determine its structure and dynamics. This hopefully provides a useful account on how seismology is used as a tool to probe a specific part of the Earth's interior, its inner core, to both a seismological and a non-seismological readership. The book thus might serve as a supplementary text to general geophysics or seismology books in various graduate or advanced undergraduate courses.

There was occasionally a need to present the basic physical and mathematical principles guiding seismologists in their observation and interpretation of physical properties and dynamics of the Earth's inner core despite the existence of more specialised books on seismology. Most mathematics is moved to appendices to allow a smoother presentation in the main part of the book, and most of it is limited to the derivation of main principles used heavily in the papers on the inner core not yet derived elsewhere. I attempted to give a comprehensive review of all data and observations that have played a significant role in shaping contemporary conceptual frameworks about the inner core. Some of these datasets are included in electronic form as supplementary material to the book.

Acknowledgements

I would like to acknowledge the contributions of many workers on the Earth's inner core who have influenced this book, even if they feel that their views were not necessarily represented in the most accurate or detailed way. I am grateful to Susan Francis from Cambridge University Press, who first approached me with the idea of proposing and writing a book on the inner core, and to her colleagues Zoë Pruce, Emma Kiddle, and Kirsten Bot for their continuing support. I would like to thank all members of the Seismology and Mathematical Geophysics Group at the Research School of Earth Sciences, whose continued patience while I was working on this manuscript was highly appreciated. Singling out my graduate students working exclusively on the inner core would not be entirely fair, as all of them shared the load during my extended absence. However, I owe special gratitude to my former PhD student Mallory K. Young, who helped me polish the original manuscript. Not only did she contribute to the clarity of the manuscript due to her love of well-worded sentences, but she provided invaluable scientific comments. I am indebted to my former PhD student Surya Pachhai for helping me prepare selected figures on normal modes and for the many discussions we had together on the subject. I am grateful to Rhys Hawkins and Drew Whitehouse of ANU Vizlab for their assistance in visualising seismological data. Special thanks to all colleagues in the D-Earth group of the Japan Agency for Marine-Earth Science and Technology, particularly to my host Dr Satoru Tanaka, for providing facilities and hospitality during the preparation of the manuscript.

I am grateful to my academic mentors: the faculty of the Geophysical Institute Andrija Mohorovičić of the University of Zagreb, which introduced me to seismology; to my PhD advisor Barbara Romanowicz, who taught me the craft of science; and to my colleagues at the Research School of Earth Sciences, Brian L. N. Kennett and Malcolm Sambridge, for their wisdom, continuing support, and invaluable academic insights.

Last but not least, I am grateful to my family for their enthusiasm and patience and, above all, to Mum and Dad, who raised me to love nature and science. Writing this book has been a true journey for me, and without their love and support, surely I would have not reached my destination.

1

On the History of Inner Core Discovery

"…and this leads us to the perception that the Earth consists of an iron core with a diameter of approximately 10 million metres, enclosed by a rocky mantle with a thickness of $1\frac{1}{2}$ million metres."[2]

Emil Wiechert (Wiechert, 1897)

"It is, however, by no means certain that a regular increase of the elastic constants to the Earth's centre is to be looked for; on the contrary, a sudden change is to be looked for where the wave paths leave the outer stony shell to enter the central metallic core which may reasonably be supposed to exist."

Richard D. Oldham (Oldham, 1900)

"We must therefore examine the possibility that the Oldham–Gutenberg discontinuity is also the outer boundary of the metallic core."

"There seems to be no reason to deny that the Earth's metallic core is truly fluid."

Harold Jeffreys (Jeffreys, 1926b)

"However, the interpretation seems possible, and the assumption of the existence of an inner core is, at least, not contradicted by the observations; these are, perhaps, more easily explained on this assumption."

Inge Lehmann (Lehmann, 1936)

"The first results for the properties of the inner core were naturally approximate. Much has been written about it, but the last word has probably not yet been said."

Inge Lehmann (Lehmann, 1987)

---

[2] Translation credit: Sebastian Rost

1.1 Early Days of Modern Science

Halley (1686) was likely the first to mention the Earth's core within the context of natural philosophy as a 'nucleus' or 'inner globe', detached from the Earth's external shell. According to Halley's model, the 'nucleus' or 'inner globe' of the Earth is a solid innermost sphere with two moveable magnetic poles which is detached from, and rotates differentially with respect to, the rest of the planet (Figure 1.1). Its existence was invoked to explain apparent observations of four magnetic poles, which were later understood to be evidence of the spatial variation of a magnetic field containing a superposition of both dipole and non-dipole components. In one of the model's variations, the Moon-sized inner globe is separated from the Earth's outer, 800 km thick shell by a liquid layer that has the same axis of diurnal rotation and the centre of gravity as the inner core (IC).

This 'Hollow Earth' model was driven by erroneous estimates of the Earth–Moon density ratio (Newton, 1687), but the idea persisted in literature for centuries. At the time in which it was published, this model was a paradigm shift that introduced some revolutionary concepts such as the existence of 'a planet within a planet', a time-varying magnetic field originating from the Earth's centre, and a 'differential rotation' of planetary shells – all of which are phenomena still lacking complete explanation.

The first estimates of the gravitational constant (5.48 g cm^{-3} by Cavendish, 1798) and Earth's mean density (5.46 $-$ 5.52 g cm^{-3} by Poynting, 1891) exceeded those of the rocks found at the Earth's surface, which prompted speculation about a denser deep Earth's interior. The idea about a molten interior was gradually replaced in the nineteenth century by the belief that the interior is solid (see, for example, a historical review by Brush, 1980) and that density must increase gradually with depth.

1.2 IC Discovery in the Context of Seismology of the Early Twentieth Century

The Earth can classically be divided into four main shells: the crust, the mantle, the outer core (OC), and the IC. There are, therefore, three main discontinuities: the crust–mantle boundary (a.k.a. Moho), the core–mantle boundary (CMB), and the inner core boundary (ICB). The term 'boundary' is used interchangeably throughout this chapter with the term 'seismic discontinuity', or just 'discontinuity', which marks a significant change in Earth's elastic properties with depth. The magnitude of that change will determine whether the discontinuity is deemed major or minor. The reader is referred to some of the classic books on seismology and the Earth's interior to find out more about the definition of discontinuities in elastic properties (e.g. Stein and Wysession, 2003; Shearer, 2009).

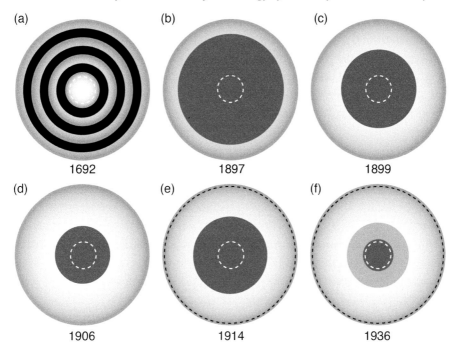

Figure 1.1 Conceptual representation (in chronological order) of early models of the Earth's interior featuring inner shells (represented by circles or rings), with published values for their radii scaled to the radius of the outer circle (Earth's surface). The white dashed circle in the centre represents the modern day IC. The black dashed circle near the surface in the last two models is the Moho. (a) Halley's 'Hollow Earth' model from 1686, based on philosophical considerations of Earth and Moon density estimates. The thickness of the outer three solid (light grey rings) and hollow shells (black rings) are 1/8 of the Earth's radius. The diameter of the 'inner globe', as Halley termed it, is 1/4 that of the Earth's diameter. (b) Wiechert's hypothetical model of the Earth from 1897, based on theoretical considerations of Earth's density. The solid core has a radius of about 4/5 of the Earth's radius. (c) Oldham's hypothetical model of the Earth from 1899 based on theoretical considerations of Earth's density and observations of earthquake waves at large distances. The solid core has a radius of about 0.55 that of the Earth's radius. (d) Oldham's model of the Earth from 1906 adjusted based on seismological observations. The solid core had a radius of about 0.4 that of the Earth's radius. (e) Gutenberg's model of the Earth from 1914, based on seismological observations and consequent analysis. The solid core had a radius of about 0.54 that of the Earth's radius. (f) Lehmann's model of the Earth from 1936, based on seismological observations and consequent analysis. The solid inner core (IC) had a radius of 1405 km. The liquid outer core (OC) had a radius close to the modern day values (2900 km).

Non-seismologists often think about probing the Earth in terms of seismic profiling with controlled sources where reflection and refraction principles can be used. However, probing the Earth's deepest shells using high-quality earthquake waves (passive seismology) was not easy, particularly in the early years of seismology when the data available were few and of low quality. It is important to realise that the Earth's radial profile is far from the simple, divided model described in the paragraph above.

Seismologists had to be innovative and develop different, often less direct, ways to detect and characterise discontinuities. For example, the Wiechert–Herglotz method of the early twentieth century was the first to determine radial profiles of velocity in the Earth (reference). The derivation of the Wiechert–Herglotz method is given in Fowler (2005) and Lowrie (2007). The profiles initially revealed a blurred and limited version of Earth. The restriction that velocity must increase with depth had to be relaxed and the number of observations had to be increased to construct more reliable travel time curves of various seismic phases.

At the turn of the nineteenth century, Wiechert suggested that the Earth's interior could be subdivided into two shells: a metallic core and a rocky mantle (Wiechert, 1897). Each had a constant density, which radically differed from the existing view that density gradually increased with depth. Wiechert performed a quantitative analysis using existing data for the principal moments of inertia and Earth's ellipticity (degree of flattening), and estimated a density of 8.2 kg m^{-3} for the metallic core and 3.2 kg m^{-3} for the mantle. His work yielded an estimate for the ratio between the radius of the core (α') and the radius of the Earth (α): $\alpha'/\alpha = 0.78$, where Earth's radius was considered 6378.2 km.

Oldham (1900) expressed a similar view about Earth's structure (see his statement at the beginning of this chapter), and without referring to Wiechert's earlier work, he concluded that the radius of the iron core is 0.55α based on hypothetical minimums for the variation of density. Oldham reported teleseismic observations of two distinctive body wave phases (P and S waves, at that time referred to as condensational and distortional waves) as well as surface waves. His empirical travel time curves reveal a regularly decreasing curvature towards larger epicentral distances and a lack of observations of seismic phases for epicentral distances beyond 90°. Assuming that the ray paths are part of a circular arc, he calculated that the epicentral distance of 90° corresponds to a bottoming depth of about 3000 km. He suggested that this depth coincided with the radius of the iron core, but did not establish a clear argument for why this would be the case. Brush (1980) suggests this may be why Oldham was not credited with discovering the core–mantle boundary (CMB) despite the fact that his CMB depth estimate was close to modern day values.

As the number of recorded teleseismic earthquakes grew, Oldham (1906) measured more P and S waves. He noted that P waves propagate more slowly in the core. He also observed a seismic shadow zone on the side of the Earth opposite the earthquake, where no P waves were recorded. He attributed their absence to P-wave refraction along the CMB. Due to misinterpretation of seismic waves at large epicentral distances (e.g. surface reflecting mantle SS waves were mistaken for S waves penetrating through the core), he concluded that the radius of the core is 0.40α, but he was unable to recognise that the core was liquid. Despite these deficiencies, the official discovery of the Earth's core is attributed to Oldham, who followed up with a number of papers addressing the nature of the core.

Mohorovičić (1910) used differential calculus and ray theory to solve for the depth of the discontinuity between the crust and the mantle. His approach combined geophysical forward and inverse methods. A critical finding that underlined his rigorous mathematical approach and subsequent discovery was the observation of two distinct arrivals of P and S waves on seismograms at a specific epicentral distance range. The observation required a sudden change in elastic properties at a depth of 54 km, which he interpreted as the depth of separation between the Earth's crust and mantle. This value agrees well with modern estimates for the study area (northeast of Croatia) (e.g. Stipčević et al., 2011).

Gutenberg (1914) estimated the CMB to reside at a depth that is not very far from today's figure of about 2889 km (Kennett et al., 1995). Jeffreys (1926b) observed an S wave shadow zone beginning at an epicentral distance of about 103.8° from the earthquake. This result indicated that the core was molten, since shear waves do not propagate through liquids. Interestingly, Stjepan Mohorovičić (the son of Andrija Mohorovičić), makes the following remark in his 1927 paper: "At the depth of 6000 km, there will be a dominance of the heaviest and noblest metals, predominantly gold and platinum, thus we can call this core Ptau. Maybe it would have been more correct to only call this hypothetic core 'Core'."[3] (Mohorovičić, 1927).

These discoveries and the state of knowledge established a solid observational basis in the first quarter of the twentieth century to accompany advances in theoretical seismology, then a relatively new discipline, and technological advances in analogue instrumentation for recording ground motion associated with earthquakes.

1.2.1 Lehmann's Discovery of the IC

In the framework of Halley's philosophy, Lehmann's discovery (Lehmann, 1936) of the Earth's IC, as she termed it, could be thought of as proof of the existence of an inner globe, separated from the outer shells of the planet by the liquid OC.

---

[3] Translation credit: Christian Sipl.

Lehmann observed arrivals of compressional waves, termed P', at angular distances from earthquakes not predicted by an entirely liquid core model. She accommodated her observations by modelling the Earth's interior with a smaller solid core inside the liquid core. In her model, earthquake compressional waves travel faster through the IC than through the OC, so that apart from transmission, a reflection occurs when the waves reach the IC. Due to insufficient data, Lehmann assumed the compressional velocity of the IC and OC to be 8.6 and 8.0 km/s, which are lower than modern day estimates (e.g. Dziewoński and Anderson, 1981; Kennett et al., 1995), and as a result obtained an IC with a radius of about 1405 km. The theoretical predictions matched the available observations at the time, but as more data gradually became available, the radius of the IC was adjusted to 1221 km. The 'IC' became a modern term and was used in the new edition of travel time curves published in 1939 (Jeffreys, 1939).

Birch (1940) suggested that the IC was solidifying from the OC and that the inner-core boundary (ICB) was a phase transition. Bullen (1946) suggested that the IC was solid. It was recognised that the solidification results in latent heat release (Verhoogen, 1961) and compositional buoyancy (Braginsky, 1963), which drive convection in the OC.

1.3 Confirmation of the Discovery and Early Seismological Studies

Two important objectives were established once the IC hypothesis was proposed. The first was to confirm its boundary within the OC and the second was to confirm its solidity. Surprisingly, it took more than three decades to achieve this confirmation, and less surprisingly, it happened during the Cold War. The need of the world powers to discriminate between nuclear explosions and earthquakes drove significant progress in observational seismology at that time. Non-proliferation seismology was in its infancy during the 1960s, resulting in unprecedented seismic arrays comparable to large radio antennas. The most remarkable undertaking was the construction of the Large Aperture Seismic Array (LASA) in Montana, which consisted of more than 500 instruments distributed over an area of about 100 km in radius. Each of 21 clusters consisted of 25 elements that were distributed in a regular shape. By amplifying the signal and cancelling out microseismic noise (see Section 2.4.2), it was possible to detect signals, for instance, from the underground nuclear explosions in Nevada.

It was not until 1971 that a seismological study of Earth's free oscillations (Dziewoński and Gilbert, 1971) produced evidence for the solidity of the IC, which will be discussed in more detail in Chapter 3. This was accompanied by a number of papers presenting supposed observations of the PKJKP phase (compressional waves that convert to shear waves at the ICB, propagate through the IC, and convert back to the compressional waves as they exit the IC) (e.g. Julian et al., 1972); more details on this will be given in Chapter 3.

2

Seismological Tools to Study the Inner Core

"The goal of seismology is to study the interior of the Earth, and to continue where the geologist stops; it has in modern seismographs a sort of binoculars that enables us to look into the largest of depths."[4]

Andrija Mohorovičić (Mohorovičić, 1913)

2.1 Introduction to Seismic Waves

Seismic sources generate a disturbance in the Earth's interior known as seismic waves. When these waves reach any given point within the Earth's interior or surface, that point can be thought of as a new source of a spherical wave – a principle known as the Huygens principle. In a similar manner to light waves, seismic waves can reflect, refract, and become diffracted at the Earth's structural discontinuities. Therefore, the disturbance in the Earth's interior, resulting from a source forcefully emitting seismic waves, can be considered a disturbance to a seismic wavefield (e.g. Kennett, 2001, 2002; Aki and Richards, 2002). This wavefield has one component that propagates through the Earth and along its boundaries in two ways: body waves and surface waves, respectively (e.g. Jeffreys, 1976; Lay and Wallace, 1995; Stein and Wysession, 2003; Chapman, 2004; Shearer, 2009; Kennett, 2009). The other component is stationary and a result of interference among intersecting waves. This part of the wavefield is commonly known as *normal modes* or *free oscillations* of the Earth (e.g. Dahlen and Tromp, 1998). Seismologists who research the Earth's IC keep both of these seismic wavefield components in their toolboxes.

2.2 Body Waves in an Elastic, Isotropic Medium

According to the theory of infinitesimal strain, the stress and strain tensors in an elastic medium are related through the following relationship, known as Hooke's law:

---

[4] Translation credit: Marijan Herak.

$$\sigma_{ij} = \mathbf{C}_{ijkl}\epsilon_{kl}, \tag{2.1}$$

where σ_{ij} and ϵ_{kl} are the second-order tensors of stress and strain, each consisting of nine elements, and $\mathbf{C}_{ijkl}$ is the fourth-order tensor consisting of 81 elements. The values of i and j are 1, 2, and 3, each representing an independent direction in Cartesian coordinates. However, since the stress and strain tensors only have only six independent elements, the number of elements in $\mathbf{C}_{ijkl}$ can be reduced to 36. Furthermore, for an isotropic medium in which material properties are independent of the direction or orientation of the sample, the number of elastic moduli can conveniently be reduced to only two. These two elastic moduli are called Lamé constants. Equation 2.1 can be then rewritten as

$$\sigma_{ij} = \lambda \mathbf{\nabla} \cdot \mathbf{u}\, \delta_{ij} + 2\mu\epsilon_{ij}, \tag{2.2}$$

where λ and μ are Lamé constants, δ_{ij} is the Kronecker delta function, $\mathbf{u}$ is the vector of displacement, and its divergence, $\mathbf{\nabla} \cdot \mathbf{u}$, is a scalar representing the dilatation (or compression) of the volume. While λ is not straightforward to describe, μ is referred to as the shear modulus or rigidity, i.e. the resistance of an elastic medium to shear (transverse) deformation. Both constants are measured in pressure units $(\mathrm{N\ m^{-2}})$.

The medium experiences local oscillations as waves pass through, but the particles in the medium do not travel with the wave. In an elastic and perfectly isotropic medium, we can relate the displacement vector to the strain tensor, and then by means of Hooke's law and Newton's second law, we can derive two differential wave equations that represent two different types of motion: compressional and shear waves. Compressional (primary or P waves) are a dilatational or compressional disturbance, $\mathbf{\nabla} \cdot \mathbf{u}$, travelling through a perfectly elastic, isotropic medium according to

$$\frac{\partial^2}{\partial t^2}(\mathbf{\nabla} \cdot \mathbf{u}) = \frac{\lambda + 2\mu}{\rho} \mathbf{\nabla}^2(\mathbf{\nabla} \cdot \mathbf{u}). \tag{2.3}$$

In the above equation, the dilatation (compression) of the volume, $\mathbf{\nabla} \cdot \mathbf{u}$, travels with the compressional velocity $\sqrt{\frac{\lambda + 2\mu}{\rho}}$, where $\mathbf{u}$ is the displacement vector, ρ is density, and λ and μ are Lamé constants. Ground particles move in the direction of wave propagation.

Shear (secondary or S waves) will present a rotational disturbance, $\mathbf{\nabla} \times \mathbf{u}$, which travels through an elastic, isotropic medium according to

$$\frac{\partial^2}{\partial t^2}(\mathbf{\nabla} \times \mathbf{u}) = \frac{\mu}{\rho} \mathbf{\nabla}^2(\mathbf{\nabla} \times \mathbf{u}). \tag{2.4}$$

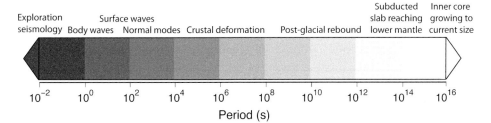

Figure 2.1 Typical periods of different types of seismic waves used in seismology compared to periods of major geologic events.

In Equation 2.4, the rotational disturbance, $\nabla \times \mathbf{u}$, is a vector that travels with shear velocity $\sqrt{\frac{\mu}{\rho}}$, where $\mathbf{u}$ is the displacement vector, ρ is density, and μ is rigidity. Ground particles move in the direction perpendicular to wave propagation. When the medium is liquid or gas, rigidity is near zero since there is no resistance to shear deformation for most practical purposes. In this case, Equation 2.4 expresses the inability of the rotational disturbance to propagate; this has been observationally confirmed, e.g. shear waves do not propagate through the ocean water (or the liquid OC).

The disturbance caused by seismic waves is manifested through a wide range of frequencies or periods, as shown in Figure 2.1. In exploration seismology, low periods (i.e. high frequencies) are used to investigate near-surface and crustal properties, whereas in regional and global seismology, typical dominant periods of body waves are of ≈ 1 s. Surface waves have dominant periods of several tens of seconds, with a period of 20 s being the nominal period used to determine surface wave magnitude. Free oscillations are studied at periods of hundreds and tens of thousands of seconds. Seismic noise spans a wide range of periods, with longer periods resulting from the interaction of the solid Earth with ocean waves and shorter periods usually resulting from anthropogenic causes. Atmospheric phenomena also contribute to seismic noise near a period of 1 s. Geological phenomena on the planetary scale, including crustal deformation due to tectonic forces and IC growth due to the planet's cooling, occur in much longer cycles.

2.3 Ray Paths and Travel Time Curves

In a high-frequency approximation (when periods approach zero and frequency go to infinity), the seismic wavefield properties are neglected entirely and the waves propagating from one point to another can be represented by a single ray (e.g. Červený, 2005). For many practical purposes, e.g. travel time measurements in seismological studies of the IC, this is a valid approximation because the accuracy of measurements is sufficiently high to describe the studied phenomenon. However, one should keep in mind that the ray theory becomes cumbersome when complex phenomena, e.g. diffraction along the inner core boundary (ICB), are encountered. Seismic body waves follow similar physical laws as do waves of light, and in a high-frequency approximation they can be thought of as rays. Since the elastic properties of the Earth generally increase with depth (i.e. higher density and seismic velocity), these rays are not straight, but curved. The longer the epicentral distance, the deeper they penetrate within the planet.

The travel time curves are travel times of different seismic waves (also referred to as *phases*) shown as a function of epicentral distance (the angular distance between the earthquake source approximated by a point and the seismic recorder). Travel time curves were first constructed towards the beginning of the twentieth century and are gradually evolving as numerous travel time data become available from an accumulating number of earthquakes and instruments installed at new locations worldwide. The first empirical travel time curves were simple, and, as we saw in Chapter 1, they played a crucial role in adding inner shells to the spherically symmetric models of the Earth. Figure 2.2 shows travel time curves for phases sensitive to the IC. The phases that are directly sensitive to the IC are shown in black; other relevant phases used either in conjunction with or those that interfere with the IC sensitive phases are shown in grey.

Although Inge Lehmann referred to the core-sensitive waves as P′, they are most commonly called PKP, because they consist of three legs: 1) P wave leg in the mantle travelling from the source to the CMB (abbreviated as P), 2) refracted P wave travelling through the core (abbreviated as K)[5], and 3) P wave leg in the mantle travelling from the CMB to the receiver (Figure 2.3a).

As we can see from Figure 2.2a, the PKP travel time curves consist of several branches: PKP(AB), PKP(BC), PKP(CD) (also referred to as the PKiKP phase, which reflects from the ICB), and PKP(DF) (referred to widely as the PKIKP phase). The PKP curve is characterised by a triplication caused by the particular ray path geometry and the distribution of P wave velocity as a function of depth. A graphical representation of these phases through the Earth is shown in Figure 2.3a for an epicentral distance of 153°, where all PKP phases theoretically exist. Figure 2.3b further illustrates the geometry of PKIKP ray paths for the range of epicentral distances between 120° and 180°. We can see that for the PKIKP rays leaving the earthquake focus, the closer the ray path is to the vertical, the greater the epicentral distance and depth of the bottoming point (the steeper, the deeper). This is to be expected since the P wave velocity increases at the transition from the OC to the IC. Travel time increases with epicentral distance, so the slope of PKIKP is positive.

---

[5] Note: In the early seismological days in Europe, the German language was common among seismologists and some of the terms became rooted in the seismological vocabulary. One such word was 'der kern', which is German for 'the core', and it is represented by the letter K in PKP.

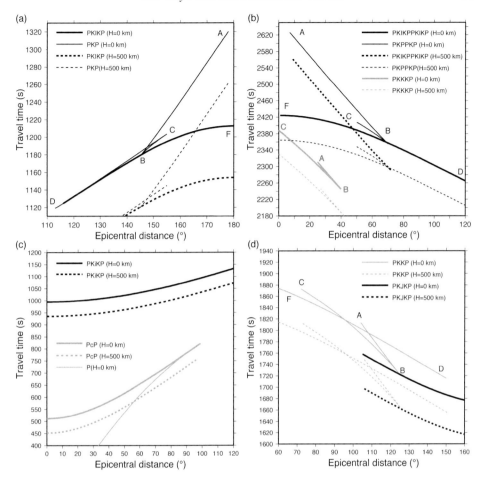

Figure 2.2 Theoretical travel time curves for IC-sensitive phases (black) and related phases (grey) from model ak135 (Kennett et al., 1995). Solid curves are computed assuming a source at the surface of the Earth. Dashed curves correspond to a source at a depth of 500 km. Capital letters indicate different branches of a particular phase. (a) PKP travel time curves, with PKIKP shown as the thickest curve. (b) PKPPKP and PKKKP travel time curves, with PKIKPPKIKP shown as the thickest curve. (c) PKiKP, PcP and P travel time curves. (d) PKJKP and PKKP travel time curves.

Whereas PKIKP traverses the IC, the particular ray path geometry and properties of the inner shells of the Earth allow for another phase, PKP(BC), to exist in a relatively narrow range of epicentral distance ($\approx 145°–155°$). However, unlike PKIKP waves, PKP(BC) waves bottom in the lowermost part of the OC, according to 'the steeper, the deeper' principle. When the epicentral distance reaches 145.5°, PKP(BC) becomes diffracted around the ICB. Thus, PKP(BC) does not penetrate the IC deeper than 352.5 km beneath the ICB (Figure 2.4).

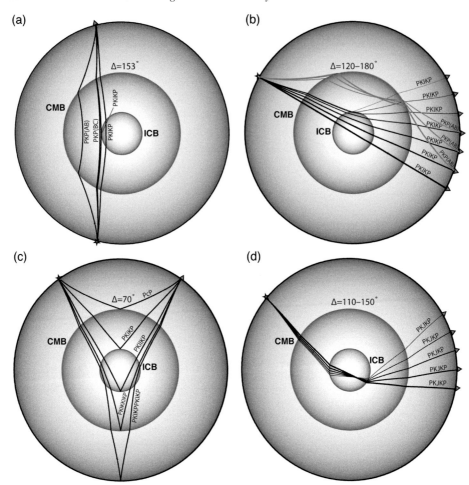

Figure 2.3 Cross-sections of the Earth along ray paths of major IC-sensitive phases. Three major shells and boundaries (ICB and CMB) are shown. The source location is shown by a star, and the receiver location is shown by a triangle. Both source and receiver are assumed to be at the Earth's surface. (a) PKP waves at an epicentral distance of 153°. (b) PKIKP (black) for the epicentral distance range 120°–180° at each 10° increment, with line-thickness increasing with epicentral distance. PKP(AB) (grey) for epicentral distances of 150°, 160°, and 170°. (c) PcP, PKiKP, PKIIKP, PKIKKIKP and PKIKPPKIKP at an epicentral distance of 70°. (d) PKJKP waves for an epicentral distance range of 110°–150°, with line-thickness increasing with epicentral distance.

PKP(AB) phases traverse the middle part of the OC. The graphical representation of the PKP(AB) ray paths through the Earth, shown in Figure 2.3b, reveals that the larger the epicentral distance, the longer the travel times; therefore, the PKP(AB) curve is positive (to make the figure clearer, we only show PKP(AB) for epicentral distances of 150°, 160°, and 170°). Note that the rays that leave the earthquake focus closer to the vertical reach smaller epicentral distances, which

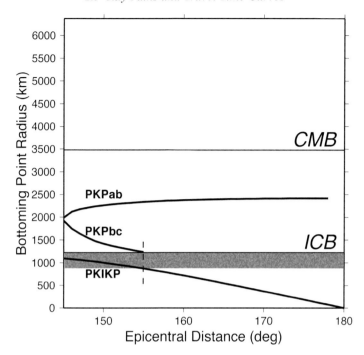

Figure 2.4 Radii of bottoming points of PKIKP, PKP(BC), and PKP(AB) waves as a function of epicentral distance (thick black lines). The CMB and ICB are illustrated by horizontal lines and separate major Earth layers. For a seismic source located at the Earth's surface, the PKP(BC) phase is not predicted beyond an epicentral distance of 155.5° (marked by a dashed line). At that epicentral distance, PKIKP waves bottom in the IC approximately 352.5 km below the ICB. The upper IC layer corresponding to PKP(BC)-PKIKP data is shown in grey. After Tkalčić (2015).

is opposite from PKIKP and PKP(BC) behaviour. It is evident from a comparison with the PKIKP phase that for a given change in epicentral distance, the change in the incident angle is larger for PKP(AB). This explains the steeper slope of the AB branch in Figure 2.2a.

A PKP phase that traverses the core, reflects from the bottom side of the Earth's free surface, and then travels one more time through the core is referred to as PKPPKP (Figure 2.3c). According to the same nomenclature, we can distinguish PKIKPPKIKP from PKPPKP(BC), PKPPKP(CD) and PKPPKP(AB), which are also referred to P′P′ waves. The corresponding travel time curve, which is opposite to that of PKP, is shown in Figure 2.2c. As the epicentral distance decreases, the travel time decreases as well, resulting in a negative slope.

Phases with multiple reverberations along the lower side of the CMB are commonly used to study OC structure. The simplest is referred to as PKKP, where the double K denotes the two legs in the OC, and reflects once off the inner side of the CMB. During the Cold War, Berkeley Seismographic Station recorded up to

13 multiples (P13KP) (Bob Uhrhammer, personal communication). These phases are sometimes referred to as the 'whispering gallery' of phases. Apart from PKKP, another higher order phase is PKKKP, where the triple K marks three legs in the OC (two reflections from the inner side of the CMB). PKPPKP arrives theoretically in the coda of PKKKP and corresponding depth phases for short epicentral distances and the travel time curves of PKPPKP and PKKKP are of similar slopes (see Figure 2.2b).

PKiKP waves are reflections from the outer ICB. This phase is expected on seismograms for small epicentral distances (Figure 2.2c, Figure 2.3c) up to the PKP triplication distance (Figure 2.2a, Figure 2.3a), although the amplitude progressively decreases at the larger epicentral distances. PKiKP have been used in conjunction with PcP to study ICB properties and the shape of the IC.

Other phases that sample the IC, such as PKIIKP and PKIKKIKP (Figure 2.3c), are more difficult to observe. The PKJKP phase is a P wave converted to an S wave at the ICB, and has been notoriously difficult to observe. Several observations were claimed, but much of the seismological community remains doubtful. The PKJKP ray paths for an epicentral distance range of $110°–150°$ and travel time curves are shown in Figures 2.3d and 2.2d. Unfortunately for the PKJKP 'hunters', potential PKJKP onset signals are obscured by the onsets of the PKKP phase, which can be rather prominent, especially if the S wave velocity in the IC is slower than predicted.

SKS are shear waves that convert to P waves at the CMB; however, they are not commonly used as they are not well observed on vertical components of seismograms. Horizontal components could be used, but since S waves are generally observed at longer periods and are highly affected by heterogeneous mantle structure, this phase is not commonly explored in IC seismology.

2.4 Travel Time Measurements

We saw in Section 2.3 that the theoretical travel times curves of IC-sensitive phases can be calculated from a spherically symmetric model of the Earth. Using these theoretical travel time curves, it is possible to identify arrivals of IC phases on seismograms recorded at different epicentral distance from the hypocentre. PKIKP waves are readily observed after large earthquakes. However, while reading absolute P wave arrival times can be a routine practice, this is not the case for PKIKP and other PKP phases. Triplication and the fact that these phases are recorded after significant attenuation and masking by both microseismic and event-generated noise makes identification non-trivial. Nonetheless, the International Seismological Centre (a successor of the International Seismological Summary) has been collecting and processing the absolute travel times, which are available in an online bulletin.[6]

---

[6] www.isc.ac.uk/iscbulletin

PKP travel time data play a significant and important role in many studies and discoveries. However, let us keep in mind that these data are contributed by seismological agencies around the world and collected by a wide range of technicians. Apart from the fact that the identification of PKP onsets will depend on an individual who analyses a seismogram, seismograph frequency responses across the world are not homogeneous. The frequency dependence (dispersion) of the B caustic is visible at low frequencies, while scattered CMB precursors to PKIKP waves are visible at higher frequencies. Both of these frequency-dependent theoretical arrivals interfere with PKIKP and PKiKP picks. Together with multi-paths at triplication distances (between 145° and 152°), they have made bulletin picks virtually useless between 125° and 152°. Travel time data in this epicentral distance range were consequently excluded from the bulletin data used to build standard Earth models like the Preliminary Reference Earth Model (PREM) (Dziewoński and Anderson, 1981).

Absolute travel time data are useful for studying Earth's structure, but they have their limitations. Earth has a heterogeneous structure, particularly its upper mantle, which can map onto the travel time measurements and interpretations based on absolute travel times. Fortunately, there is a remedy: using differential (as opposed to absolute) travel times of seismic phases with similar ray paths can reduce biases stemming from mislocations in space and time and unwanted effects imposed by near-source and near-receiver structure. Since the source and receiver effects will have a similar influence on both rays, the residual travel time anomaly can be ascribed exclusively to deep Earth structure. Early studies relied on this approach to assess deep Earth structure (e.g. Cormier and Choy, 1986; Sylvander and Souriau, 1996), although it has since been argued that smaller-scale, deeper mantle heterogeneity can independently and significantly influence each ray path (e.g. Bréger et al., 1999, 2000; Tkalčić, 2010).

Box 2.1 Definition of PKP travel time residuals

PKIKP absolute travel time residuals

$$\Delta t = PKIKP_{meas.} - PKIKP_{ak135}, \qquad (1)$$

where *meas.* denotes the measured (observed) onset, and *ak135* denotes the theoretically predicted onset of the PKIKP phase on a seismogram (in time units).[7]

PKP(AB)−PKIKP differential travel time residuals

$$\Delta t = [PKP(AB) - PKIKP]_{meas.} - [PKP(AB) - PKIKP]_{ak135}, \qquad (2)$$

---

[7] Note: ak135 is the Earth reference model (Kennett et al., 1995).

Box 2.1 (cont.)

where *meas.* denotes the measured (observed) difference,[8] and *ak135* denotes the theoretically predicted difference in onset times of PKP(AB) and PKIKP phases on a seismogram.

PKP(BC)–PKIKP differential travel time residuals

$$\Delta t = [PKP(BC) - PKIKP]_{meas.} - [PKP(BC) - PKIKP]_{ak135}, \quad (3)$$

where *meas.* denotes the measured (observed) difference, and *ak135* denotes the theoretically predicted difference in onset times of PKP(BC) and PKIKP phases on a seismogram.

Near-source and near-receiver structures can be assumed to similarly affect PKIKP and PKP(BC) ray paths, but not necessarily the PKP(AB) and PKIKP ray paths as PKP(AB) and PKIKP traverse the lowermost mantle on entirely different routes (Figure 2.3). Therefore, PKP(BC)–PKIKP measurements are a preferred and generally accepted tool to investigate IC structure. We saw in Section 2.3 that PKIKP bottoms at a radius of 865.0 km, which is 352.5 km below the ICB in model ak135 (Kennett et al., 1995). Therefore, PKP(BC)–PKIKP differential travel times technique only yields information about the upper 352.5 km of the IC. This depth range could be extended with a differential travel times technique if PKP(BC) diffracted or PKP(AB) are considered in conjunction with PKIKP waves. However, the uncertainties imposed by the ICB topography effects on the PKP(BC) diffracted phase and mantle heterogeneity effects on the PKP(AB) phase are likely greater than the benefits of using the differential travel time technique itself. Using absolute PKIKP travel times corrected for heterogeneous mantle structure is arguably the better approach to delving deeper into the IC.

Differential travel times are not simple subtractions of one phase arrival time from another. This is particularly true for PKP phases, because PKIKP is an attenuated version of PKP(BC), and, mathematically speaking, PKP(AB) is a Hilbert-transform version of PKP(BC) (Figure 2.5). Any study using absolute travel times of IC-sensitive phases should acknowledge its limitations and inherent errors. For example, for the PKP(AB)–PKIKP pair, if the absolute arrival times are simply subtracted one from another, the error would be on the order of 1 to 2 s, which amounts to the typical correction applied to PKP travel times to account for mantle heterogeneity.

---

[8] Note: In waveform cross-correlation measurements, a Hilbert transform is applied to PKP(AB) before cross-correlation with PKIKP (Figure 2.5).

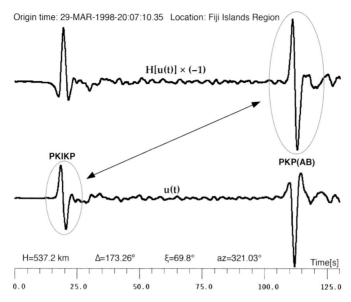

Figure 2.5 Illustration of the first step in measuring differential travel times. The bottom trace is an original seismogram taken from the TAM station dataset (see Figure 4.1) showing clear arrivals of the PKIKP and PKP(AB) phases. The top trace is a Hilbert-transformed version of the bottom trace, and multiplied by -1. Note the similarity of the PKIKP and Hilbert-transformed PKP(AB) phases (encircled), which are now ready for cross-correlation. After Tkalčić (2001).

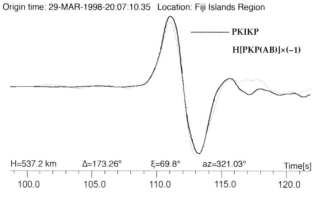

Figure 2.6 Illustration of the second step in measuring differential travel times using waveform cross-correlation. The thick solid line is the original seismogram (see Figure 2.5), and the thin solid line is the Hilbert-transformed seismogram multiplied by -1 and shifted by time δt until the best waveform fit between the PKIKP and PKP(AB) phases is obtained. After Tkalčić (2001).

The phase change a PKP(AB) wave experiences as a result of an internal caustic in the Earth is mathematically equivalent to applying a Hilbert transform operator to the original PKP(AB) waveform (Figure 2.5). The Hilbert-transformed

PKP(AB) becomes similar in shape to PKIKP, and the cross-correlation of the two phases results in a high-quality measurement of the time shift (Figure 2.6). Typically, the errors are on the order of 0.1 to 0.2 s, but the precision can sometimes be significantly reduced by noise or Earth structure so that the Hilbert-transformed PKP(AB) no longer resembles PKIKP. In addition, significant IC attenuation can broaden the pulse of the PKIKP phase. PKP(BC)−PKIKP measurements are simpler, and we can skip the Hilbert transform step because PKP(BC) does not undergo a phase change. If unaccounted for, the effect of IC attenuation on PKIKP is still an issue, however.

2.4.1 Ray Path Coverage

Here, we examine the present-day volumetric coverage of the upper 350 km of the IC by PKIKP ray paths associated with seismograms on which both PKIKP and PKP(BC) arrivals were observed, thus allowing the PKP(BC)−PKIKP differential travel time technique to be utilised. Figure 2.7 illustrates the volumetric coverage (often overlapping) of three such data sets from the last two and a half decades. In the Tkalčić et al. (2002) data set (Figure 2.7a), the differential travel times are measured using cross correlation and complemented with the highest-quality waveform correlation measurements from McSweeney et al. (1997), Tanaka and Hamaguchi (1997), and Souriau and Romanowicz (1997) to utilise limited-access data. In general, the Circum-Pacific belt is covered well, but there are vast poorly sampled volumes under the oceans. The Garcia et al. (2006) data set for deep events (Figure 2.7b) resulted from an innovative non-linear inversion algorithm, SAWIB. A limited number of new paths is introduced, but most data for deep events are included in the data set shown in Figure 2.7a. Irving and Deuss (2011) (Figure 2.7c) independently compiled cross-correlation-based measurements; the data are more recent but the ray path coverage remains a subset of previous coverage.

The Souriau and Romanowicz (1997) data set (Figure 2.7d) was used to study the relationship between IC attenuation and velocity anisotropy. Only very good (quality A) and good (quality B) data are shown, while poor (quality C) and very poor (quality D) data are omitted. The Garcia et al. (2006) data set for shallow events (Figure 2.7e) clearly shows a dramatic improvement over previous data sets, especially under Europe, Australia, and the Pacific Ocean. This is due to the inclusion of a number of shallow events in Africa, North America, and mid-ocean ridges (red stars). Despite this promising improvement, the shallow earthquake data are of a lower quality and show a much larger scatter than the deep earthquake data of Garcia et al. (2006). Finally, Figure 2.7f illustrates the combined IC coverage with the exception of shallow events shown in Figure 2.7e. This composite map of

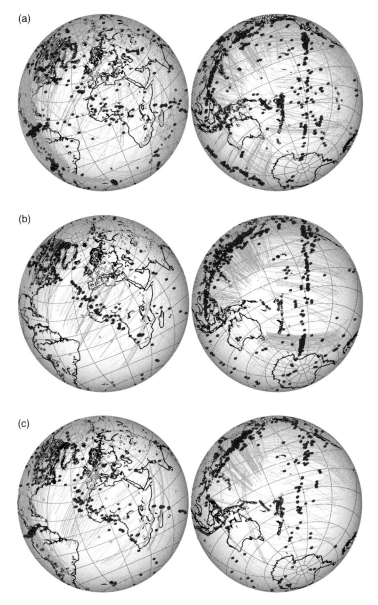

Figure 2.7 Surface projections of PKIKP ray paths through the IC are shown in orange on tilted perspective projections of the Earth centred on the western (left column) and the eastern (right column) hemisphere. Dark blue ellipses indicate the positions of the PKIKP ray bottoming points. Locations of earthquake sources are shown with red stars. Stations that recorded PKP(BC) and PKIKP waves used to compile the PKP(BC)−PKIKP differential travel time data are shown with yellow triangles. Data sets by (a) Tkalčić et al. (2002), (b) Garcia et al. (2006), and (c) Irving and Deuss (2011). (d) Souriau and Romanowicz (1997), (e) Garcia et al. (2006), restricted to shallow events only, and (f) combination of all data sets, with the exception of shallow events shown in (e). After Tkalčić (2015). (A black and white version of this figure will appear in some formats. For the colour version, please refer to the plate section.)

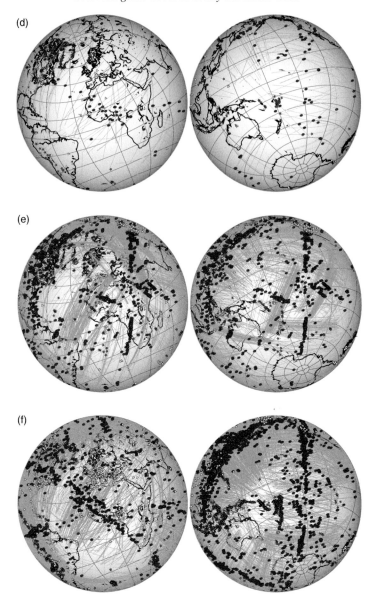

Figure 2.7 (*cont.*)

data coverage demonstrates a potential for future imaging studies of the IC using tomographic techniques, though projecting the ray paths to the Earth's surface may potentially present an overly optimistic portrayal of IC coverage and cross-pathing (see Chapter 6, Section 6.1).

2.4.2 Seismic Array Analysis

Apart from individual seismic stations, arrays of receivers have been operational from the second half of the twentieth century. Different array designs and apertures facilitate different studies of IC structure (e.g. Doornbos and Husebye, 1972; Vidale and Earle, 2000; Poupinet and Kennett, 2004; Yee et al., 2014; Wang et al., 2015; Waszek and Deuss, 2015; Tanaka and Tkalčić, 2015).

Excellent small-aperture arrays (10–20 km) have been built and used by the International Monitoring System of the Comprehensive Nuclear-Test-Ban Treaty Organisation (e.g. Warramunga Array in Northern Territory, Australia (WRA), Eliason Array in Alaska, USA (ILAR) and Yellowknife Array in the Northwest Territories, Canada (YKA); see Figure 2.8a,b,d). And following the success of the first continental-scale transportable array, SKIPPY in Australia, USArray in the United States crossed North America recently. Between the small aperture and continental arrays are regional-scale transportable arrays such as WOMBAT in Australia (and its subarrays, e.g. SEAL2; see Figure 2.8e) and the PASSCAL arrays installed around the globe. In addition, permanent national networks exist, such as NORSAR (Norway) and F-net and Hi-net (Japan), which have been used in global studies apart from seismicity monitoring purposes.

Box 2.2 Warramunga seismic and infrasound array

Warramunga Seismic and Infrasound Research Station (WRA) was established in 1968 as part of the International Monitoring System (IMS) for the Comprehensive Nuclear-Test-Ban Treaty Organisation (CTBTO) based at the United Nations Office in Vienna, Austria. WRA is located about 35 km south-southeast of Tennant Creek, Northern Territory, Australia. Today's design incorporates a 24-element seismic array and an 8-element infrasound array. The Australian National University's Research School of Earth Sciences, Seismology and Mathematical Geophysics Group, operates WRA on behalf of CTBTO. The data from WRA have been used for the monitoring purposes and numerous seismological studies of Earth's structure and dynamics (examples).

Three spiral arrays were recently deployed in Australia. The spiral array PSAR (Pilbara Seismic Array) was deployed in Western Australia by Geoscience Australia to enhance tsunami warnings. SQspa (Figure 2.8c) and WAspa, spiral arrays in southern Queensland and Western Australia, respectively, were deployed by the ANU Seismology and Mathematical Geophysics Group and Research School of Earth Sciences (RSES) as a part of the Australian Research Council Discovery Project DP130101473 (Tkalčić, H., Kennett, B. L. N., and Tanaka, S.).

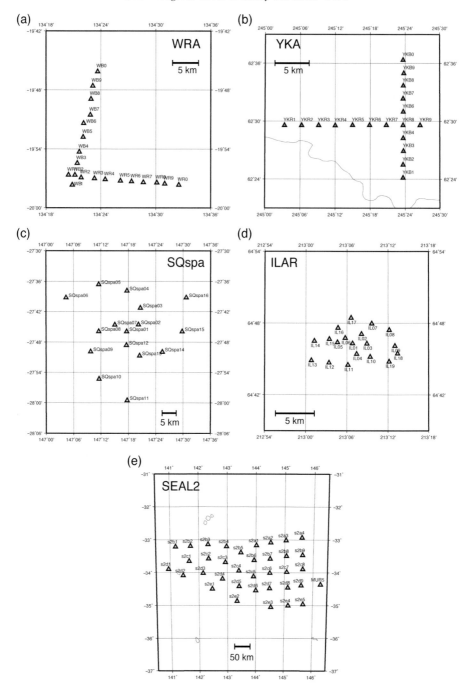

Figure 2.8 Geographic locations and designs of various arrays. Triangles mark individual array element locations. The horizontal bar in each map shows the length scale in km. (a) Warramunga Array (blue and red arm only), (b) Yellowknife Array, (c) Southern Queensland Spiral Array, (d) Eliason Array, (e) SEAL2 Array (subcomponent of WOMBAT Array).

fk analysis

In order to illustrate how small aperture arrays enhance and identify signals from the deep Earth, let us briefly describe frequency–wave number analysis (*fk analysis*), which is one of the array techniques widely used to identify the direction of the incoming energy recorded by an array. The direction of incoming energy can be important when identifying seismic phases that are not readily observed, or well-known phases accompanied by precursors and postcursors. In its simplest form, fk analysis assumes that the wave front is planar, which is sufficiently true for small aperture arrays. An fk analysis yields power spectral density as a function of the direction of the incoming energy. In order to briefly illustrate the technique, we must introduce 'slowness' first. The slowness, u, could be thought of as the inverse of the apparent velocity of a wavefront across an array (Rost and Thomas, 2002) and is defined as

$$u = \frac{sin\, i}{v_0},$$ (2.5)

where v_0 is the velocity at which the wavefront propagates beneath the array, and i is the incident angle with respect to the vertical. When incidence is vertical, slowness is zero. Conversely, the apparent velocity of the waveform across the array will be infinite, since for a vertical incidence the plane wave hits all elements of the array at the same time. The 3D slowness vector is uniquely described in spherical coordinates by the magnitude of the horizontal slowness $\mathbf{u_0}$ (magnitude of a projected 3D vector of slowness onto the Earth's surface) and back azimuth θ. The relationship between $\mathbf{u_0}$ and θ is

$$\mathbf{u_0} = \frac{1}{v_0}(\sin \theta, \cos \theta).$$ (2.6)

The fk analysis approach searches over the two quantities ($\mathbf{u_0}$ and θ) by shifting the phases of the recorded seismograms in the frequency domain and identifying the direction in space that yields the maximum energy. As the wavefront sweeps across the array, it can be shown that the total energy recorded by the array is given by

$$E = \frac{1}{2\pi} \int_{-\infty}^{\infty} |S(\omega)|^2 \left| \frac{1}{N} \sum_{n-1}^{N} e^{i(\mathbf{k}-\mathbf{k_0})\mathbf{r_n}} \right|^2 d\omega,$$ (2.7)

where $S(\omega)$ is the Fourier transform of the recorded seismogram (time series), $|S(\omega)|^2$ is the power spectral density, ω is angular frequency $2\pi f$, and f is frequency in Hz. N is the number of elements, $\mathbf{k_0}$ is the wave number corresponding to the slowness $\mathbf{u_0}$ (defined as $\omega \mathbf{u_0}$), $\mathbf{k}$ is the wave number corresponding to a signal with slowness $\mathbf{u}$, and $\mathbf{r_n}$ is the location vector of the nth element relative to the array reference point. The relative power in decibels is plotted as a function of horizontal slowness and as the distance from the origin, and back azimuth θ is plotted as the angle from geographic north.

The second factor under the integral in Equation 2.7 is known as the array response function (ARF), which is controlled by the array design, including station spacing, aperture, and shape. Figure 2.9 reveals that while the small aperture array (WRA) has a good response at low periods (high frequencies), the opposite is true for the regional array (SEAL2). This is expected, as the station spacing of SEAL2 is about 20 times larger than that of WRA.

Small aperture arrays have good amplification properties, somewhat limited resolution in slowness and back azimuth, and the design can play a significant role in Earth structure recovery. They are readily used by the International Monitoring System (IMS) of the Comprehensive Nuclear-Test-Ban Treaty; for example, Figure 2.10 shows ARFs for two regular IMS arrays: WRA and YKA. For these arrays, the diversity of azimuthal orientation of the inter-stations vectors is limited, and this often presents a problem for Earth structure recovery (Kennett et al., 2015). Figure 2.11 shows ARFs for two arrays with untraditional designs: ILAR, a small aperture array comparable to WRA and YKA, and SQspa (Tkalčić, 2015), which is almost twice as large in aperture. The most obvious advantage of irregular arrays is their enhanced detection capacity and minimised cost and resource requirements (Kennett et al., 2015). The spiral arm design of SQspa allows for a broader spread of instruments (and therefore better slowness and back azimuth resolution), while keeping the number of elements relatively low. The energy side lobes are efficiently suppressed for a spiral arm array due to the versatility of the resulting inter station vectors (Kennett et al., 2015).

Dense seismic arrays such as the Hi-net array in Japan, WOMBAT array in Australia, or USArray in the United States (and their subarrays) are currently being utilised to increase the signal-to-noise ratio of weak core-sensitive seismic phases and form part of a multi-array probing of Earth's heterogeneity (Stipčević et al., revised for Geophysical Journal International, 2016). An example of observation of an exotic seismic phase, podal PKPPKP waves, and their precursors, is shown in Figure 2.12. Although the observed slowness and back azimuth might vary from their theoretical values, it is evident that the main signal and precursors are coming from the same direction in space and that the incidence is steep. It is probably that the signal comes from a PKPPKP wave, with discontinuities in the upper mantle causing the back-scattering (Tkalčić et al., 2006).

2.5 Normal Modes and Measurements in Frequency Domain

As mentioned before, to infer Earth's internal velocity structure from normal modes is like guessing the internal structure of a custom-made pipe organ from the noise

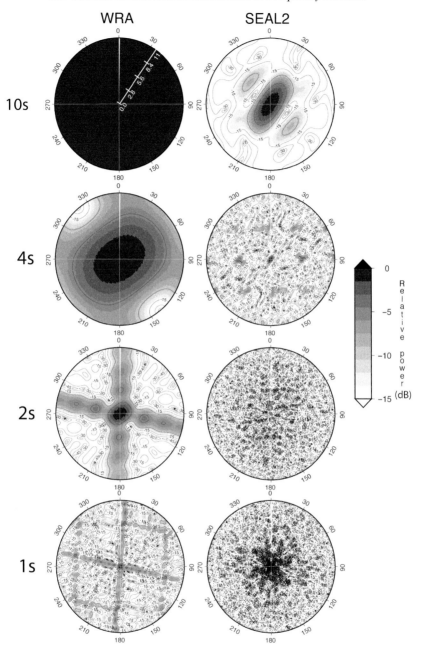

Figure 2.9 Array response functions (ARF) calculated in the frequency domain for the small-aperture WRA array (left) and regional SEAL2 array (right). For the array locations, see Figure 2.8a,e. The relative power in decibels is plotted as a function of horizontal slowness along the radial axis (in $s/°$) with 0 in the centre, and back azimuth θ forming an angle from geographic north. ARFs are shown in the slowness range $0 - 11.2\,s/°$ for a monochromatic wave entering the array domain vertically from below. From top to bottom, the period (frequency) of the monochromatic wave decreases (increases).

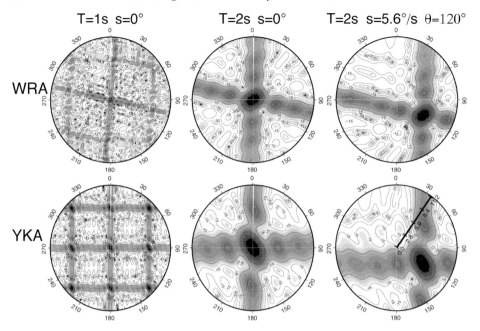

Figure 2.10 A comparison of ARFs for WRA (top) and YKA (bottom), calculated in the frequency domain for a monochromatic wave with slowness $s = 0.0$ and period equal to $1.0\,s$ (left) and $0.5\,s$ (middle). On the right: monochromatic wave of $0.5s$ with slowness $5.6\,s/°$ and back azimuth $120°$. The magnitude of ARF is expressed as relative power in decibels on the same scale as in Figure 2.9. The scale in the lower right diagram shows the slowness range.

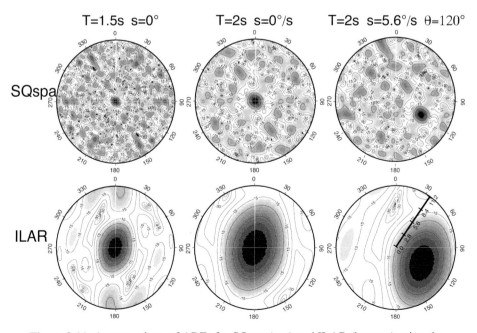

Figure 2.11 A comparison of ARFs for SQspa (top) and ILAR (bottom) using the same parameters and relative power scale as in Figure 2.10.

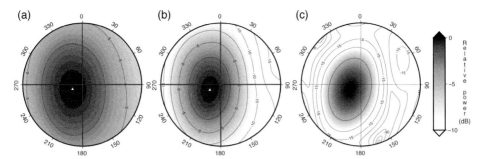

Figure 2.12 Observation of podal PKPPKP waves (Figure 2.3) and their precursors on the ILAR array for the event reported in Tkalčić et al. (2006). A strong signal is observed around slowness $0.22 \, s/°$ and back azimuth $\theta = 244°$ for: (a) the main phase and (b) the precursor. (c) ARF for the slowness and back azimuth parameters observed in (a) and a monochromatic wave of 2 s. (A black and white version of this figure will appear in some formats. For the colour version, please refer to the plate section.)

it makes when thrown down the stairs. It would require multiple tosses, each time with a slightly different stairway and number and type of stairs. Each time, a unique combination of pipes will be activated, enabling us eventually to deduce how they are distributed within the pipe organ. A topple down a stairway is a metaphor for an earthquake, and the sound that a pipe organ makes is a metaphor for the excitation of Earth's normal modes. It takes a number of earthquakes to observe a variety of normal modes.

In an elastic, isotropic, spherically symmetric, and non-rotating Earth, normal modes can take two forms: spheroidal ($_nS_l$), in which a particle undergoes both radial and tangential motion, and torsional or toroidal ($_nT_l$), where a particle executes motion on a spherical surface (no radial motion). n is the radial order, and l is the angular order (Dahlen and Tromp, 1998). n represents the number of crossings of zero amplitude of motion (a.k.a. nodes) along the Earth's radius, and l determines the pattern of deformation on the Earth's surface. When $n = 0$, the mode is referred to as the fundamental mode, and nodes of $n > 0$ are overtones.

Any wave motion within the Earth can be expressed as a sum of normal modes. Unlike body waves, whose motion can be visualised in a high-frequency approximation using rays, normal modes present a pattern of motion that samples the entire volume of the Earth simultaneously. Spheroidal modes could be thought of as a combination of compressional and vertically polarised shear waves; therefore, they can describe Rayleigh wave motion. Toroidal modes comprise of horizontally polarised shear waves and can describe Love wave motion only, as they do not contain a radial component of motion).

It is often said that the Earth oscillates like a bell after large earthquakes, but how do we isolate a particular mode of oscillation? How do we identify a signature of an IC-sensitive normal mode in a recorded seismogram? What would be an equivalent procedure using normal modes to measuring travel times of IC-sensitive phases in the time domain? The answers to these questions are not straightforward. The simple solution would be to apply a narrow bandpass filter to the recorded seismogram around the theoretically predicted frequency of a particular normal mode. We could then apply a Fast Fourier Transform to switch to the frequency domain, measure the values of the excited frequencies, and compare them with theoretical values. In practice, however, there are several other steps that need to be implemented before we can proceed with this type of normal mode analysis.

Box 2.3 Basic steps in preprocessing of data for normal mode analysis (a case of spheroidal modes)

  (i) Request continuous recordings from vertical components of long period seismograms for about 10 days after the origin time;

 (ii) Remove instrument response (by deconvolution);

(iii) Remove tidal effects by applying a high-pass filter (e.g. MacDonald and Ness, 1961);

(iv) Remove time windows containing aftershocks, other events, glitches and clipped segments;

 (v) Discard the first 10–15 hours after the origin time for most IC-sensitive modes;

(vi) For each record, cut a window with a minimum length matching that of the mode's Q-cycle (typically about 60–70 hours for IC-sensitive modes);

(vii) Discard noisy time-series;

(viii) Apply a Hanning taper to each window;

(ix) Apply a Fast Fourier Transform to each window to convert time series to complex spectra in the frequency domain.

Figure 2.13 displays the vertical component of a long period displacement seismogram after one of the largest (moment magnitude equal to 8.2) and deepest (focal depth equal to 647 km) earthquakes in recorded history – the 1994 Bolivia earthquake. The selected recording station is located on the Horn of Africa, in Djibouti (ATD). The instrument response and long-period tidal signal have been removed by deconvolution. We can see from a visual inspection of this seismogram that approximately 5–10 hours after the origin time, the ground motion gradually reduces back to lower amplitude level characteristic of the time interval before the passage of earthquake waves. However, the level of ground displacement is still

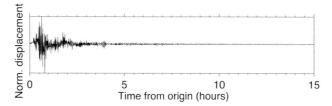

Figure 2.13 The first 15 hours of broadband displacement data after the 1994 Bolivia earthquake, recorded on the vertical component of station ATD in Djibouti. The vertical axis is normalised with respect to maximum displacement.

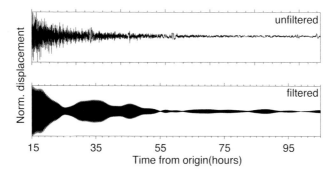

Figure 2.14 The next 90 hours of broadband displacement data recorded on the vertical component of station ATD in Djibouti. The vertical axis is normalised with respect to the maximum displacement. The seismogram starts 15 hours after the origin time of the 1994 Bolivia earthquake. The instrument response and tidal effects were removed. (Bottom) same as above, but bandpass filtered (4.82–4.87 mHz) around the theoretical frequency of the IC-sensitive $_{13}S_2$ multiplet.

greater than that caused by microseismic and signal-generated noise alone. We can see from Figure 2.14 that the exponential decay continues beyond 15 hours after the origin time and that the seismogram is characterised by various other types of motions, *i.e.* aftershocks and brief high-frequency signals. The bottom seismogram in Figure 2.14 shows the same recording, but after the application of a narrow bandpass filter around the theoretical frequency of the IC-sensitive mode, $_{13}S_2$. Box 2.3 shows the steps in preprocessing long period data for normal mode analysis.

2.6 Sensitivity of Normal Modes

Normal modes are described by *eigenfrequencies* and *eigenfunctions*, which represent the solutions of the wave equation and depend on the radius, shape, and properties of the Earth. A frequency perturbation of a normal mode, $\delta\omega$ due to a perturbation in the bulk and shear modulus of elasticity and density can be expressed as

$$\delta\omega = \int_0^a \left(\delta k\, K_k + \delta\mu\, K_\mu + \delta\rho\, K_\rho\right) dr, \tag{2.8}$$

where a is the normalised radius of the Earth, K_k, K_μ, and K_ρ are the sensitivity kernels for bulk modulus (k), shear modulus (μ), and density (ρ), respectively as a function of radius (r). Since normal modes of certain frequencies sample the entire volume of the Earth, it is instructive to visualise the sensitivity kernels. The equations are expressed in terms of eigenfunctions as follows:

$$K_k = \frac{1}{2\omega}\left(r\dot{U} + 2U - kV\right)^2, \tag{2.9}$$

$$K_\mu = \frac{1}{6\omega}\left(2r\dot{U} - 2U + kV\right)^2 + \left(r\dot{V} - 2V + kU\right)^2 + \left(r\dot{W} - W\right)^2$$
$$+ \left(k^2 - 2\right)\left(V^2 + W^2\right). \tag{2.10}$$

P wave velocity sensitivity is given by

$$K_\alpha = 2\rho\alpha K_k. \tag{2.11}$$

S wave velocity sensitivity is given by

$$K_\beta = 2\rho\beta\left(K_\mu - \frac{4}{3}K_k\right). \tag{2.12}$$

In the above equations, U, V, and W are radial eigenfunctions, which can be obtained using the MINOS software (Woodhouse, 1988). Figure 2.15 shows the sensitivity of the IC-sensitive modes up to radial order 15 to both P (solid lines) and S wave velocity (dashed lines) from the spherically symmetric model PREM (Dziewoński and Anderson, 1981). The sensitivity kernels are calculated from Equations 2.11 and 2.12. As we can see, most normal modes sensitive to IC P wave velocity sample the upper third of the IC. There are a few normal modes with maximum sensitivity in the IC that sample the upper half. The modes in the figure are not normalised with respect to each other, but with respect to their relative sensitivity along the entire Earth's radius. There is near-zero sensitivity to P wave velocity in the lower third of the IC. Quite a few normal modes are sensitive to S wave velocity, with their sensitivity being more equally spread along the radius of the IC. Among the featured modes, there are two modes with pronounced sensitivity to S wave velocity near the Earth's centre.

Figures 2.16 and 2.17 show the P and S wave velocity sensitivity profiles of selected IC-sensitive modes: most (except the last two in Figure 2.17) were previously shown in Figure 2.15. In general, the sensitivity is larger in the Earth's core (mostly in the OC) than in the mantle. The lateral variation of normal mode eigenfunctions is expressed in terms of spherical harmonics, which are the solutions to Laplace's equation on the sphere, as follows:

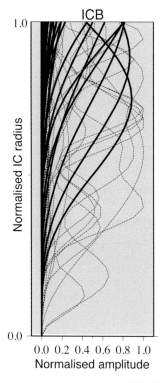

Figure 2.15 Sensitivity to P wave velocity (solid line) and S wave velocity (dashed line) as a function of normalised IC radius for all normal modes containing energy in the IC up to radial order 15. Each mode is expressed in terms of energy density, which is computed for the anisotropic PREM model (Dziewoński and Anderson, 1981). The amplitude of each mode is normalised with respect to the maximum amplitude of that mode.

$$\nabla^2 \Phi(\theta, \phi) = 0, \qquad (2.13)$$

where θ is colatitude and ϕ is longitude. A normalised spherical harmonic of degree l and order m can then be expressed as

$$Y_l^m(\theta, \phi) = (-1)^m \left[\frac{(2l+1)}{4\pi} \frac{(l-m)!}{(l+m)!} \right]^{\frac{1}{2}} P_l^m(cos\theta)exp(im\phi), \qquad (2.14)$$

where P_l^m is the associated Legendre polynomial of the appropriate degree (e.g. Press, 2007). The harmonic degree l and order m represent the nodal structure of the spherical harmonic Y_l^m. The l value is defined by the number of zero crossings on the surface, while the $|m|$ value is defined by the number of zero crossings that pass through the pole. The value of m lies between $-l$ and l ($2l+1$ in total). An arbitrary twice-differentiable scalar function f on the sphere may be written in terms of spherical harmonics and the coefficients, C_l^m, as follows:

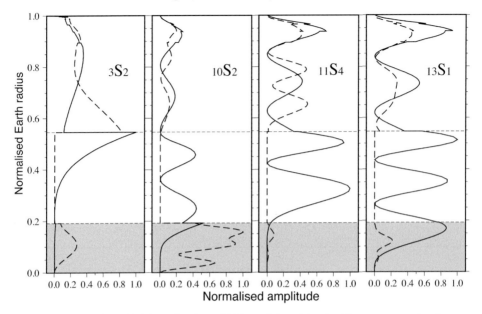

Figure 2.16 Sensitivities of selected IC models with significant energy in the IC to P wave velocity (solid line) and S wave velocity (dashed line) as a function of normalised radius. Each mode is expressed in terms of energy density, which is computed for the anisotropic model PREM (Dziewoński and Anderson, 1981). Dashed lines are CMB and ICB discontinuities. Grey area indicates the IC. (a) $_3S_2$; (b) $_{10}S_2$; (c) $_{11}S_4$; (d) $_{13}S_1$.

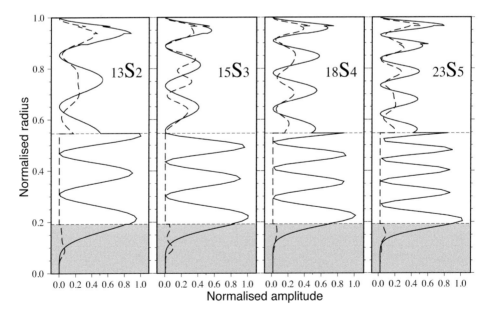

Figure 2.17 Same as Figure 2.16. (a) $_{13}S_2$; (b) $_{15}S_3$; (c) $_{18}S_4$; (d) $_{23}S_5$.

$$f(\theta, \phi) = \sum_{l=0}^{\inf} \sum_{m=-l}^{l} C_l^m Y_l^m(\theta, \phi). \tag{2.15}$$

The above expression is useful in terms of expressing *splitting functions*, which is the topic of the next section.

2.7 Splitting of Normal Modes

The Earth is a closed-boundary system and, therefore, permits standing waves, or normal modes. Different modes that share the same frequency are referred to as *degenerate* modes. So far, we have explained normal modes in the context of a spherically symmetric Earth; however, the Earth's interior is far from a stack of homogeneous, spherical shells. Most significantly, the Earth rotates, is not a perfect sphere, and is exposed to the tidal forces of the Moon and Sun, and the material of which the Earth is composed is not perfectly isotropic. These imperfections break the degeneracy of eigenfrequencies. The frequency of each isolated *multiplet* (a mode that does not interact with other modes) is split into a number of frequencies or resonances, referred to as *singlets*. This phenomenon is known as *normal mode splitting*. Masters and Gilbert (1981) were the first to observe anomalous splitting for a mode sensitive to the IC ($_{10}S_2$; Figure 2.16b).

An isolated multiplet with an angular order l is split into $2l + 1$ singlets. If only rotation and ellipticity cause the splitting, the frequency of mth singlet, ω_m, is defined as

$$\omega_m = \omega_0(1 + a + bm + cm^2), \tag{2.16}$$

where ω_0 is the central frequency of the multiplet, b is the Coriolis force, and a and c account for the Earth's ellipticity and second-order effects of rotation. Therefore, provided that we can measure the frequencies of individual singlets, we are equipped with an indirect measure of the strength of the phenomena that cause the splitting. A displacement seismogram for an isolated normal mode (Fig. 2.18) is calculated using the normal mode summation of singlets (Woodhouse and Girnius, 1982), which can be expressed as

$$u_j(t) = \sum_{m=-l}^{l} R_{jm} exp\left[i\left(H_{mm'} + \mathbf{I}\omega_0\right)t\right]a_{m'(0)}, \tag{2.17}$$

where $u_j(t)$ is the displacement at station j, and R is the *receiver vector* matrix. Each row of R is a vector of $2l + 1$ elements and represents a product of the displacement eigenfunction at the Earth's surface and the spherical harmonics that describe the lateral variation of normal modes in a spherically-symmetric Earth according to Equation 2.14. The azimuthal order m' is the same as m but in the

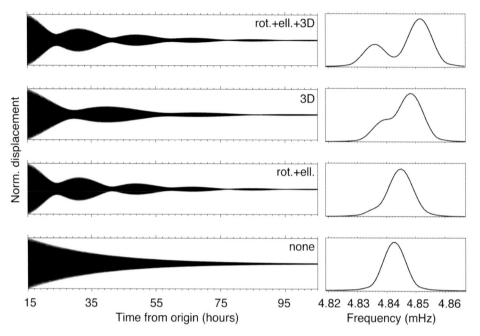

Figure 2.18 Theoretical (synthetic) displacement seismograms (vertical components), computed using Equation 2.16 for the $_{13}S_2$ multiplet. Various combinations of effects (departures from spherical symmetry) were taken into account when comparing with the observed seismogram shown in Figure 2.14. From top to bottom: 3D effects from published c_s^t coefficients (He and Tromp, 1996), effects of Earth's ellipticity and rotation; only 3D effects; only ellipticity and rotation effects; no effects (spherically symmetric Earth model). Vertical axes are normalised with respect to maximum displacement.

complex conjugate and transpose domain. ω_0 is the central frequency of a multiplet. **I** is an identity matrix, and $a(0)$ is the *source excitation vector* consisting of $2l + 1$ elements. H is the complex splitting matrix of dimension $(2l + 1) \times (2l + 1)$ for an isolated multiplet, and includes information about the rotation, ellipticity, and 3D structure of the Earth sensed by a normal mode (Woodhouse and Dahlen, 1978). The complex splitting matrix, $H_{mm'}$, can be expressed as

$$H_{mm'} = \left(a + bm + cm^2\right)\delta_{mm'} + \sum_{s=0,even}^{s} \sum_{t=-s}^{s} \gamma_{ls}^{mm't} c_s^t. \qquad (2.18)$$

The first term in Equation (2.18) includes the effects of Earth's rotation and ellipticity (Equation 2.16). Since the rotation and ellipticity of the Earth are well determined, the splitting matrix can be readily corrected for those effects. The second term incorporates the effect of 3D heterogeneous structure. c_s^t are the *structure (splitting) coefficients* that depend linearly upon the Earth's heterogeneity of the harmonic degree s and order t (e.g. Masters et al., 1982; Giardini et al., 1987). $\gamma_{ls}^{mm't}$ is the Gaunt integral and is represented by an integral over three fully normalised spherical harmonics (e.g. Woodhouse and Dahlen, 1978). When the splitting matrix is diagonal, the particular mode 'sees' Earth structure as axisymmetric. If we think of splitting coefficients as perturbations from the central frequency, we can express them similarly to perturbation in Equation 2.8. Their linear relationship to lateral variations of Earth's 3D structure can be written as

$$c_s^t = \int_0^a \delta \mathbf{m}_s^t(r) \cdot \mathbf{M}_s(r) r^2 dr \,, \tag{2.19}$$

where a is the normalised radius of the Earth, $\mathbf{M}_s(r)$ is the sensitivity kernel, and $\delta \mathbf{m}_s^t(r)$ is a vector of the expansion coefficients of the radial and lateral basis functions (Earth's 3D structure). The vector $\delta \mathbf{m}_s^t(r)$ consists of perturbations in P wave velocity (α), S wave velocity (β), and density (ρ):

$$\delta \mathbf{m}_s^t(r) = \left(\frac{\delta \alpha_s^t}{\alpha}, \frac{\delta \beta_s^t}{\beta}, \frac{\delta \rho_s^t}{\rho} \right). \tag{2.20}$$

Perturbations from a 1D spherically symmetric Earth model (1D) can be readily computed using a 3D model of Earth structure (e.g. from a tomographic imaging). Once the splitting matrix is determined, Equation 2.17 can be used to compute synthetic seismograms, as illustrated in Figure 2.18. The uppermost synthetic seismogram shown in this figure is calculated considering the effects of ellipticity, rotation, and 3D heterogeneity determined from published splitting coefficients (He and Tromp, 1996). Below are two synthetic seismograms calculated using either only 3D heterogeneity effects, or only the rotation and ellipticity effects. When compared with the observed seismogram (Figure 2.14), it is evident that the effects of rotation are dominant. Finally, the bottom synthetic seismogram is calculated using $H_{mm'} = 0$, i.e. there are no rotation, ellipticity, nor 3D effects. It is clear from Equation 2.17 that in this case only a simple exponential decay term remains (representing the attenuation). Global lateral variation of the splitting of an isolated normal mode can be visualised by means of *splitting functions*. Splitting functions represent lateral variation of the radial average of the 3D structure sensed by a normal mode. Mathematically, using the formalism introduced in Equation 2.15, the splitting functions are expressed in terms of spherical harmonic basis functions and given by

$$\sigma(\theta, \phi) = \sum_{s=0,even}^{2l+1} \sum_{t=-s}^{s} c_s^t Y_s^t(\theta, \phi) \,. \tag{2.21}$$

There are several ways to compute splitting functions and Earth structure from these complex spectra, one of which is through *two-step inversion*. In a two-step inversion, the splitting functions (or splitting coefficients, c_s^t) are first computed using a non-linear inversion (e.g. Woodhouse et al., 1986; Ritzwoller et al., 1988; Giardini et al., 1988; Li et al., 1991; Resovsky and Ritzwoller, 1998; He and Tromp, 1996; Romanowicz et al., 1996) and then the splitting coefficients are linearly inverted for 3D structure (both elastic and/or anisotropic). Alternatively, a direct inversion can be performed (Durek and Romanowicz, 1999).

In the final stage of data 'preprocessing', long-period seismograms are transformed from the time domain to complex spectra in the frequency domain (Box 2.3). The next stage is 'data processing' or 'data analysis', which involves measuring the splitting coefficients. First, the *receiver strips*, $\mathbf{d}(\omega, t)$, can be derived by multiplying Equation 2.17 by the inverse of the receiver vector matrix and written as

$$\mathbf{d}(\omega, t) = \mathbf{R}^{-1}\mathbf{u} = exp\left[i\left(H + \mathbf{I}\omega_0\right)t\right]a(0). \tag{2.22}$$

One example of receiver strips for the $_{13}S_2$ mode is shown in Figure 2.19 for the 1994 Bolivia earthquake. All available global long-period seismograms were used, and all the preprocessing steps described in Box 2.3 were applied. We can see from these receiver strips that all raw data (displacement seismograms from

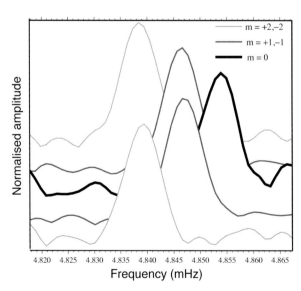

Figure 2.19 Receiver strips for $_{13}S_2$ mode derived from the 1994 Bolivia earthquake long-period seismograms and preprocessing steps described in Box 2.3.

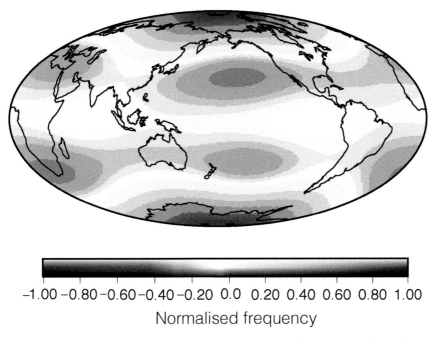

−1.00 −0.80 −0.60 −0.40 −0.20 0.0 0.20 0.40 0.60 0.80 1.00

Normalised frequency

Figure 2.20 Elastic splitting functions for the $_{13}S_2$ mode from the 1994 Bolivia and 1994 Fiji earthquakes shown as normalised frequency perturbations. (A black and white version of this figure will appear in some formats. For the colour version, please refer to the plate section.)

approximately one hundred global stations) are collapsed into only five spectra ($2l + 1 = 5$ for $_{13}S_2$ mode) sensitive to the IC. These spectra contain both the source and 3D structure information, and it is evident from Figure 2.19 that they are split, i.e. centred at different frequencies.

To eliminate earthquake source information from the receiver strips, we can use an autoregression method (Masters et al., 2000a,b) or a derivative-free approach such as the Neighbourhood Algorithm (Pachhai et al., 2016). Since the non-uniqueness of the inversion can lead to inconsistent results it is desirable to use the latter approach, which provides an approximate uncertainty measure for the structure coefficients. Figure 2.20 shows the splitting functions for the $_{13}S_2$ mode, calculated using the 1994 Bolivia and 1994 Chile earthquakes and the neighbourhood algorithm as in Pachhai et al. (2016). The splitting functions are then linearly inverted for 3D elastic/isotropic or anisotropic structure according to Equation 2.22.

3

Inner Core Surface and Its Interior

"Because the modes of group 2 would have zero energy in the inner core if it were liquid, the only way to increase their eigenfrequencies would be to increase the bulk modulus in the outer core, a condition contradicting that necessary to satisfy the data of group 1. We therefore conclude that solidity of the inner core represents the only solution consistent with the observations of normal modes."

Adam M. Dziewoński & Freeman Gilbert
(Dziewoński and Gilbert, 1971)

3.1 Introduction to Studies of the IC Surface and Its Interior

A simple glance at a tomogram derived through use of modern imaging techniques reveals a stunning variety of features within the Earth's interior. Some local and regional studies of the Earth's crust use the portion of the seismic wavefield with wavelengths of several kilometres, while other, usually global, studies concentrate on wavelengths that exceed hundreds or even thousands of kilometres. When the focus of study is shifted to larger depths, images of the deep Earth's interior are often blurry, lacking the level of detail obtained nearer the Earth's surface. Tomography often does not extend to the Earth's core. If we were equipped with a toolset similar to that used for imaging near the Earth's surface, and if we were able to apply those tools to the well-mixed OC, the images would most likely reveal homogeneous seismological structure at long wavelengths. But what about the solid IC, which sits in its centre? The IC's radius is comparable in length to IC-sensitive long wavelength waveforms and normal modes, which are sensitive to IC structure only as part of an integral over the rest of the planet's radius. Yet, we have progressed significantly in understanding the physical appearance of the Earth's IC, not just at its surface, but also deep in its interior. Questions including how the IC appears at its surface, whether it has a rough or smooth boundary, whether it contains topography, how it solidifies and at what rate, what the density contrast at the OC boundary is, and how the IC texture changes with distance from the ICB have been the driving force behind many IC studies.

In this chapter we first concentrate on studies focusing on ICB properties (Section 3.2). This includes a discussion on the density contrast at the ICB (Sections 3.3 and 3.4) and some recent attempts to characterise its topography (Sections 3.5 and 3.6). We then shift gears and see how seismology has been used to infer properties of the IC's interior, such as its aggregate state (3.7) and isotropic P-wave velocity, both in terms of its hemispherical and lateral variations (3.8). The reader is directed to Chapter 4 for a more in-depth analysis of seismic anisotropy in the IC. Section 3.9 discusses attenuation, and the relationship between attenuation and isotropic velocity is discussed in Section 3.10. Finally, in Section 3.11 we discuss recent geodynamical models that attempt to explain a range of seismological observations of both the surface and interior of the IC and enhance our understanding of Earth's mantle dynamics.

3.2 Simple View of the ICB

As mentioned in Chapter 1, Engdahl et al. (1970) reported the observation of PKiKP waves, which reflect from the ICB. The ray path of these waves in a high-frequency approximation is illustrated in Figure 2.3c. Their observation was the first confirmation that the IC had a sharply defined surface and enabled measurement of the IC radius, which was determined to be 1216 km, within several kilometres of uncertainty. The observations of PKiKP waves were used extensively in conjunction with reflections from the core–mantle boundary (CMB) to determine the IC to OC density contrast. Pioneering work by Bolt and Qamar (1970) prompted a long sequence of seismological papers on this topic.

By the late 1970s and early 1980s the boundary between the IC and OC was thought to be flat (based on observations of strong PKiKP waves) and sharply defined (based on the extremely small variation in temperature in the OC, e.g. Stevenson, 1987; Jeffreys, 1939; Engdahl et al., 1970, 1974. There is a straightforward connection between the age of the IC and its radius, which indicates how quickly the ICB moves outward as the IC solidifies. Early estimates of the IC radius and how it varies laterally (e.g. Engdahl et al., 1974) did not reveal any peculiarities in the shape of the IC, although it was suggested by Poupinet et al. (1983) that the IC might have a prolate shape in an effort to explain PKIKP travel times. There was a strong trade-off between OC radius and velocity estimates and IC radius estimates. Estimates of the IC radius gradually settled to around 1220 km (e.g. Dziewoński and Anderson, 1981; Kennett et al., 1995), though this value depends strongly on the velocity in the OC just above the ICB.

3.3 Density Contrast at the ICB from Body Waves

The density discontinuity at the ICB provides invaluable constraints on the dynamics of solidification, and it is crucial for calculations of thermal evolution within the Earth's core (e.g. Buffett et al., 1996; Nimmo et al., 2004). In particular, the ratio of density above and below the ICB is useful for magnetohydrodynamics and dynamo simulations. How do we obtain such a ratio? The ICB is the deepest confirmed first-order discontinuity within the Earth, and there is a significant trade-off between the thickness of the Earth's layers and their elastic parameters, particularly density. Kennett (1998) argues that the constraints imposed on the polynomials describing density in each depth interval in spherically symmetric models are based on mathematical convenience rather than an attempt to allow for different physical processes. Luckily, seismologists have found a way to study the density contrast directly from the amplitudes of body waves interacting with the ICB. We will describe in the following section how this has been achieved, starting with the pioneering work of Bolt and Qamar (1970).

The body wave method utilises direct measurements of amplitude ratios of PKiKP and PcP waves (see Figure 2.3c). Despite simplifications introduced on the nature of boundary conditions at the ICB and Earth structure such as homogeneous half-spaces in contact and frequency-independent attenuation, it is an elegant method that yields reasonable, empirically proven predictions. Tkalčić et al. (2009) presented expressions for reflection and transmission coefficients at solid–liquid and liquid–solid boundaries (from ground displacements) using the same set of boundary conditions and mnemonics as in the pioneering work of Bolt and Qamar (1970). For convenience, a full derivation for the liquid–solid boundary case is shown in Appendix A. The results were checked against those calculated by routines distributed with Seismological Algorithms (Doornbos, 1988) as well as against the amplitude ratios predicted by dynamic ray tracing routines (e.g. Červený, 2005).

A system of three equations with three unknowns obtained for the ICB (as in Appendix A) is shown below:

$$
\begin{bmatrix}
\beta_2(tan^2\varphi''' - 1) & 0 & 2\alpha_2 tan\varphi'' \\
\beta_2 & -\alpha_1 tan\varphi & -\alpha_2 tan\varphi'' \\
2\frac{\beta_2^3}{\alpha_1^2}tan\varphi''' & \alpha_1\frac{\rho_1}{\rho_2}sec^2\varphi & -\alpha_2\frac{\beta_2^2}{\alpha_1^2}(tan^2\varphi''' - 1)
\end{bmatrix}
\begin{bmatrix}
B/A \\
C/A \\
D/A
\end{bmatrix}
=
\begin{bmatrix}
0 \\
-\alpha_1 tan\varphi \\
\frac{\rho_1}{\rho_2}\alpha_1 sec^2\varphi
\end{bmatrix}.
$$

$$(3.1)$$

In Equation 3.1, a P wave with amplitude A is incident onto a horizontal liquid–solid boundary at an angle φ in the Cartesian co-ordinate system, with x pointing eastward and z pointing upward (see Figure A.1). C is the amplitude of the reflected

P wave (at an angle φ'), and D is the amplitude of the transmitted P wave (at an angle φ''). B is the amplitude of the refracted SV wave (at an angle φ'''). α_1 and ρ_1 are the compressional velocity and density of the liquid, while α_2, β_2, and ρ_2 are the compressional and shear velocities and density of the solid.

A similar system of equations can be obtained for the solid–liquid boundary (CMB):

$$
\begin{bmatrix}
\beta_2(1 - tan^2\varphi''') & 2\alpha_2 tan\varphi & 0 \\
\beta_2 & \alpha_2 tan\varphi & \alpha_1 tan\varphi'' \\
2\frac{\beta_2^3}{\alpha_1^2}tan\varphi''' & \alpha_2\frac{\beta_2^2}{\alpha_1^2}(tan^2\varphi''' - 1) & -\frac{\rho_1}{\rho_2}\alpha_1 sec^2\varphi''
\end{bmatrix}
\begin{bmatrix}
B/A \\
C/A \\
D/A
\end{bmatrix}
$$

$$
= \begin{bmatrix}
2\alpha_2 tan\varphi \\
\alpha_2 tan\varphi \\
-\frac{\beta_2^2}{\alpha_1^2}(tan^2\varphi''' - 1)
\end{bmatrix}.
\tag{3.2}
$$

Instead of a transmitted wave, there is a reflected SV wave with amplitude B, rotated using the right-hand co-ordinate system. The boundary conditions are the same as for the liquid–solid boundary. The derivation of this system of equations, although not shown here, follows similar steps to those shown in Appendix A for the system of equations 3.1.

The above systems of equations (3.1 and 3.2) can be solved for the transmission/reflection coefficients of any liquid–solid and solid–liquid boundary within the Earth given a known ray geometry and elastic parameters on both sides of the boundary. The effects of energy partitioning at the ICB on P-wave amplitudes can be better understood if we plot the transmission/reflection coefficients as a function of ray geometry. Figure 3.1 shows the transmission/reflection coefficients for the ICB using the elastic parameters from model ak135 (Kennett et al., 1995), plotted as a function of the incidence angle: $90° - \varphi$, where the angle of emergence φ can range between $0°$ and $90°$. A density of 12,139 kg m^{-3}, a compressional wave speed of 10.289 km/s, and a shear wave speed of 0.000 km/s were assumed representative of the liquid side of the ICB. On the solid side of the ICB, a density of 12,704 kg m^{-3}, a compressional wave speed of 11.043 km/s, and a shear wave speed of 3.504 km/s were assumed. There is a critical angle at $67.8°$. These ray-theoretical coefficients are generally independent of frequency but become frequency-dependent near grazing incidence of P or S waves on either side of the boundary at around $90°$, where diffracted/evanescent waves are excited (Richards, 1976).

In Figure 3.2 the transmission/reflection coefficients for PKiKP (left) and PcP (right) are plotted as a function of epicentral distance. For the ICB, the same model parameters were used as in Figure 3.1. On the liquid side of the CMB, a density of 9915 kg m^{-3}, a compressional speed of 8.000 km/s, and a shear wave speed of

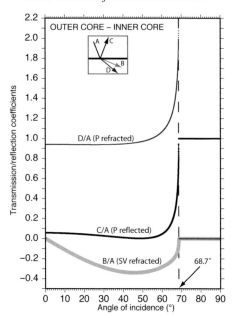

Figure 3.1 Transmission/reflection coefficients for the P wave crossing from the IC to the OC as a function of the angle of incidence. Model ak135 (Kennett et al., 1995) provides the elastic parameters in the surrounding layers. For the geometry of the rays and the notation of the transmission/reflection coefficients, see Appendix A and Equation 3.1. Modified from Tkalčić et al. (2009).

0 km/s were assumed. On the solid side of the CMB, a density of 5551 kg m^{-3}, a compressional wave speed of 13.660 km/s, and a shear wave speed of 7.281 km/s were assumed. Due to the interaction with the CMB, the portion of PKiKP energy that is refracted decreases with increasing epicentral distance. The portion of PKiKP energy that reaches the ICB (left diagram) and reflects also decreases with epicentral distance. At about 70°, almost no energy is reflected back to the OC. At the same time, the CMB-reflected portion of PcP increases significantly with epicentral distance and peaks at about 60°. The reverse behaviours of PKiKP and PcP amplitudes as a function of increasing epicentral distance illustrate the difficulty in identifying high quality observations of PKiKP and PcP from a single earthquake at the same station.

The above considerations aid understanding of the partition of energy at the major core boundaries. How do we proceed from here to determine the IC to OC density contrast? Conveniently, in the systems of equations 3.1 and 3.2, the density contrast between liquid and solid appears explicitly. PKiKP waves interact with the CMB and ICB three times, while PcP waves interact only once, with the CMB. Therefore, assuming that at near vertical incidence PcP and PKiKP waves have similar paths through the mantle, with the significant differences between their ray

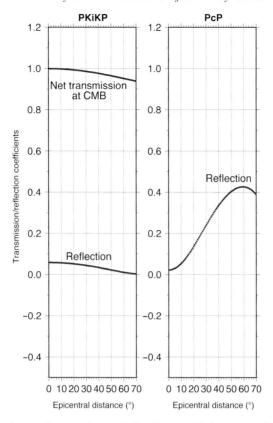

Figure 3.2 Products of transmission/reflection coefficients as a function of epicentral distance (see Equation 3.3) for PKiKP waves (left), which cross the CMB twice and reflect at the ICB and for PcP wave (right), which reflect from the CMB (right) as a function of epicentral distance (see Equation 3.3). The same values for the elastic parameters were used in Figure 3.1. After Tkalčić et al. (2009).

geometries occurring only in the OC, the expressions for transmission/reflection coefficients can yield theoretical predictions of the IC to OC density contrast as a function of epicentral distance in a spherically symmetric Earth. At short epicentral distances, PcP and PKiKP ray paths are near vertical and similar in the mantle, but they increasingly differ as epicentral distance increases (e.g. Shearer and Masters, 1990; Cao and Romanowicz, 2004; Koper and Pyle, 2004). With the assumption of nearly identical ray paths in the mantle, an expression for the PKiKP/PcP displacement amplitude ratio $\mathcal{A}(\Delta)$ as a function of epicentral distance can be written as:

$$\mathcal{A}(\Delta) = \frac{\mathcal{T}_{CMB-d}(\Delta)\mathcal{R}_{ICB}(\Delta)\mathcal{T}_{CMB-u}(\Delta)}{\mathcal{R}_{CMB}(\Delta)} \frac{\eta_Q(\Delta)}{\eta_S(\Delta)}, \qquad (3.3)$$

where $\mathcal{T}_{CMB-d}(\Delta)$ (equivalent to D/A in Equation 3.2) and $\mathcal{T}_{ICB-u}(\Delta)$ (equivalent to D/A in Equation 3.1) are the transmission coefficients of a PKiKP wave travelling down and up, respectively, through the CMB. $\mathcal{R}_{ICB}(\Delta)$ (equivalent to C/A in Equation 3.2) is the reflection coefficient of a PKiKP wave at the ICB, and $\mathcal{R}_{CMB}(\Delta)$ (equivalent to C/A in Equation 3.1) is the reflection coefficient of a PcP wave at the CMB. Equation 3.3 means that when assuming a close proximity of ray legs in the mantle, the amplitude ratio PKiKP/PcP as a function of epicentral distance is dependent on three factors: (1) the transmission/reflection coefficients at the CMB and ICB, (2) the geometrical spreading ratio, η_S, and (3) the elastic attenuation ratio, η_Q. Factors (2) and (3) stem from the ray legs of PKiKP in the OC.

In Equation 3.3, η_S is the PKiKP/PcP ratio geometrical spreading factor, defined in equation 9.69 of Aki and Richards (2002) as:

$$\eta_S(\Delta) = \sqrt{\frac{cos^2(i_{\text{PKiKP}})}{cos^2(i_{\text{PcP}})} \frac{p_{\text{PcP}}}{p_{\text{PKiKP}}} \left|\left(\frac{dp}{d\Delta}\right)_{PcP} \Big/ \left(\frac{dp}{d\Delta}\right)_{PKiKP}\right|}, \qquad (3.4)$$

where p is the ray parameter, i is the incident angle, and Δ is the epicentral distance. It is assumed that the source is on the surface.

The effect of attenuation stemming from the slightly different ray orientations in the upper mantle can be ignored at distances less than about $30°$. We can also neglect the effect of attenuation in the lowermost mantle where the ray paths of PKiKP and PcP diverge. It remains that the loss of energy of PKiKP compared to that of PcP can be explained by the difference in attenuation resulting from the additional PKiKP ray legs in the OC:

$$\mathcal{A}^{\text{PcP}} = \mathcal{A}^{\text{PKiKP}} exp(\frac{\pi \Delta t}{Qt}), \qquad (3.5)$$

where Δt is the two-way travel time of PKiKP waves in the OC, Q is the path averaged quality factor in the OC (a measure of relative energy loss per oscillation cycle), and T is the period of PKiKP waves. Although attenuation $1/Q$ in the middle of the OC can be confidently set to near zero, as seen from the high frequency content of PKnKP (e.g. Cormier and Richards, 1976), a possible zone of higher attenuation (lower Q) has been suggested between 200 and 400 km above the ICB (Zou et al., 2008).

The factor η_Q used in Equation 3.3 is the attenuation correction factor (applied to the double leg of PKiKP in the OC):

$$\eta_Q = exp(-\frac{\pi \Delta t}{Qt}), \qquad (3.6)$$

where all the symbols are the same as in Equation 3.5.

The attenuation in the OC has a small impact on the theoretical amplitude ratio. In fact, the uncertainty in amplitude ratio measurements exceeds the resolution of current OC attenuation models. The attenuation accumulated through the mantle has a larger effect, however, and geometrical spreading will have a dramatic impact on the theoretical curve. If geometrical spreading effects are disregarded, the predicted shape of the amplitude ratio as a function of epicentral distance curve would not change, though it would shift towards higher amplitude values for a given epicentral distance. This effect becomes more prominent as epicentral distance decreases (see Figure 3 of Tkalčić et al., 2009). Finally, the largest influence on the theoretical amplitude ratio comes from the combination of the four transmission/reflection coefficients appearing in Equation 3.3; therefore, it is to be expected that these ratios will be very sensitive to the elastic properties near the CMB and ICB.

Equation 3.3 predicts the PKiKP/PcP displacement amplitude ratio as a function of epicentral distance, taking into account the partition of energy at the main boundaries, geometrical spreading, and attenuation. Thus, we can construct theoretical curves of displacement amplitude ratios as a function of epicentral distance for various values of density contrast. We can compare these curves with the data if we have observations of both PcP and PKiKP waves on the same seismogram originating from the same event. The more observations we have, the more credible the comparison with the theoretical values. This comparison enables estimation of the density contrast at the ICB. An example of the theoretical predictions and a comparison with the data is shown in Figure 3.3, which is taken from Tkalčić

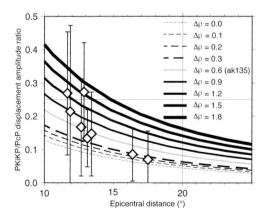

Figure 3.3 PKiKP/PcP amplitude measurements and their uncertainties (the median values are shown by diamonds, and the uncertainties are shown by error bars) plotted as a function of epicentral distance for a varying density contrast, $\Delta\rho$, at the ICB (shown by the lines of varying thickness). Theoretical values for the velocity on the both sides of the ICB are taken from model ak135 (Kennett et al., 1995). Modified from Tkalčić et al. (2009).

et al. (2009) and the associated dataset. The data points and their uncertainties (it will be discussed later how these uncertainties were obtained) lie well within the theoretical bounds, but not on a single density contrast curve.

The studies of Bolt and Qamar (1970) and Souriau and Souriau (1989), which were in general based on sparse global coverage of the ICB, predicted a higher density contrast by a factor of two or more than that predicted from normal modes: 550–600 kg m^{-3} (Dziewoński and Anderson, 1981). For example, Souriau and Souriau (1989) obtained a range of 1350–1660 kg m^{-3} from the amplitude ratios of PKiKP/PcP observed in earlier studies (Bolt and Qamar, 1970; Bolt, 1972; Buchbinder et al., 1973; Engdahl et al., 1974) and new data recorded at short epicentral distances ($\Delta < 45°$) by the Warramunga array. Even though their data indicated a much higher density contrast than that in PREM (Dziewoński and Anderson, 1981) and the normal mode study of Masters (1979), they concluded that the IC was not purely of iron, which would require an even higher density contrast. Shearer and Masters (1990) considered both body waves and normal modes, but could not find enough seismograms with both PcP and PKiKP waveforms. Body wave studies concluded that the density contrast at the ICB must be less than 1000 kg m^{-3}, while normal modes required even lower values of around 550 kg m^{-3}.

3.4 Modern Estimates of the Density Contrast at the ICB and Implications for IC Growth

The density contrast at the ICB is larger than it would be for a phase transition alone. When data were sparse, it was suggested that the observations of PKiKP at epicentral distances between 10° and 70° represent extreme situations (with arrivals likely being enhanced by the focusing effect of mantle heterogeneities), so that the density contrast estimates are actually an upper bound (Shearer and Masters, 1990). Strong PKiKP reflections from the ICB could be observed if the liquid fraction is small near the boundary or if the thickness of the mushy zone at the top of the IC is only several hundreds of metres (Loper, 1983), which is less than the wavelength of the observed PKIKP waves sensitive to IC structure (at the frequency of 1 Hz, this is on the order of about 10 km in that region).

As the global coverage of new and modern broadband instruments improved, the number of studies on the density contrast at the ICB increased. Our ability to observe PKiKP also improved as data accumulated and became freely available via the internet. For example, Figure 3.4 illustrates high-quality observations of PKiKP waves from Lop Nor nuclear site explosions, observed at stations in Asia (Tkalčić et al., 2009). The increase in the number of simultaneous observations of PKiKP and PcP waves gradually helped resolve the discrepancy between the normal mode

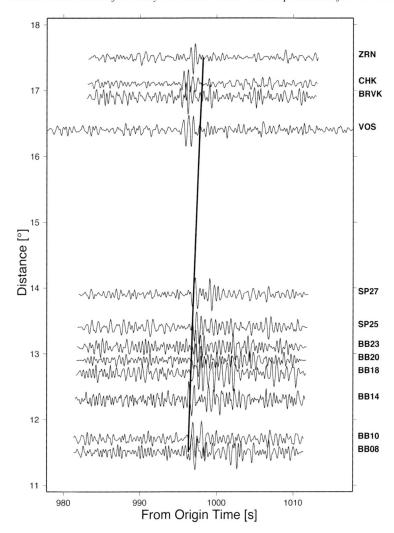

Figure 3.4 Observations of the PKiKP waves on the velocity seismograms of various stations in Asia for a Lop Nor test-site explosion from 7 October 1994. The theoretical travel time prediction of PKiKP from model ak135 (Kennett et al., 1995) is shown by a solid line. After Tkalčić et al. (2009).

and body wave studies. Masters and Gubbins (2003) recalculated the density jump from normal mode data and showed that the previous estimate based on normal modes could be raised to about 820 kg m^{-3}. Cao and Romanowicz (2004) investigated body waves and determined a density contrast of 850 kg m^{-3} based on five observations. Thus, new results from two independent datasets reconciled an old discrepancy, although other body wave studies still produced a range of different results, perhaps because the ICB sampling in each study was different. Koper and Pyle (2004) showed that uncertainty in seismic velocities near the boundary is another source of error in estimating the density contrast near the CMB. Koper and Dombrovskaya (2005) argued for lateral variation in ICB properties based on a regional variation in the PKiKP/P ratio by up to two orders of magnitude.

In the last decade there has been an increasing number of simultaneous observations of PKiKP and PcP waves, particularly those sampling the eastern hemisphere of the IC (e.g. Krasnoshchekov et al., 2005; Kawakatsu, 2006; Tkalčić et al., 2009; Tkalčić et al., 2010; Dai et al., 2012; Jiang and Zhao, 2012; Zeng and Ni, 2013). Krasnoshchekov et al. (2005) used array observations to measure absolute PKiKP amplitudes of waveforms from nuclear explosions of known location at major test sites. They normalised the strength of each source to a fixed explosive yield and found that the observed amplitude from 50° to 100° epicentral distance could not be explained by spherically symmetric ICB models. This led Krasnoshchekov et al. (2005) to suggest that the variability of the inferred ICB density contrast from 450 to 1660 kg m^{-3} indicates the existence of a mosaic of laterally variable physical properties across the ICB. Their interpretations included the possibility of strong gradients in shear modulus on either the top or bottom side of the ICB.

Motivated by the multitude of ICB density contrast results, Tkalčić et al. (2009) proposed a new approach that integrated the effects of microseismic and signal-generated noise with the amplitude measurements to account for uncertainty. They used high-quality PcP and PKiKP wave arrivals from a nuclear explosion observed at epicentral distances between 10° and 20°. The resulting uncertainties were high, precluding precise estimates of the ICB density contrast, but they did provide an upper bound of about 1100 kg m^{-3}. They observed a small ICB density contrast of 200–300 kg m^{-3} at some locations, which suggests either the existence of zones of suppressed density contrast at the ICB, a CMB density contrast stronger than 5000 kg m^{-3}, or a combination of both. In addition, an independent study showed that the amplitudes of PKiKP and PcP could be negatively correlated as a result of small-scale heterogeneity in the crust and upper mantle (Tkalčić et al., 2010). Li et al. (2014) propose a method that suppresses the effect of upper mantle scattering beneath the USArray receivers to obtain a clearer image of the ICB properties. They argue for variations in the ICB properties, with either a transitional mushy layer a few km thick or sharp horizontal changes in ICB elevation.

These later studies illustrated that ray theory has limitations: it is not ideal for precise measurements of the ICB density contrast, though it can be used to calculate upper bounds. More recent estimates of the ICB density contrast are less than 1000 kg m^{-3}, which is close to the upper bound from normal mode studies. While the discrepancy between the normal modes and body wave observations has generally been resolved, some body wave studies have yielded very low estimates for the ICB density contrast.

If normal modes also require a small density contrast of about 400 kg m^{-3} (Gubbins et al., 2008), the effect on the compressional velocity profile in the thermo-chemical boundary layer at the bottom of the OC may require a modest heat flux from a very old IC. Alternatively, the CMB density contrast could have a sporadic spatial peaks, which would also result in low PKiKP/PcP amplitude ratios. There are some indications that the ICB could be rough (e.g. Poupinet and Kennett, 2004) and that the IC just below the ICB could have an alternating pattern of solidified and less solidified texture (e.g. Cormier, 2007). If the ICB is characterised by a mosaic of variable density contrasts to which seismic body waves are sensitive, it is likely that the density fluctuations are restrained to the top of the IC. For instance, Bergman (2003) argued that the top of the IC is a dendritic mushy zone, in which coexisting interdendritic fluid pockets explain the observed nature of attenuation (e.g. Cormier et al., 1998) and attenuation of body wave anisotropy (e.g. Creager, 1992; Souriau and Romanowicz, 1996). A low ICB density contrast estimate could result if a PKiKP wave reflects from the IC at a location of particularly low density. Some authors argued that at the base of the OC there may exist a compositional change and/or large-scale temperature difference and that IC formation processes may be different between the two hemispheres (Yu et al., 2005), which would require a laterally varying density contrast at the ICB. Possibly consistent with lateral variations in the P-wave velocity gradient in the lowermost OC is a region of laterally varying viscosity associated with a frequency dependent shear modulus in the lowermost OC (Cormier, 2009). These lateral variations may be coupled to lateral variation in the solidification process of the IC (Cormier, 2007) and lateral variations of flow in the OC (Aubert et al., 2008).

3.5 ICB Radius and Topography

Jeffreys (1939) adjusted Lehmann's initial estimate of the IC radius from 1405 to 1250 km based on slowness estimates from PKIKP waves. The value of 1250 km persisted in literature until the 1970s, when several studies proposed smaller values after adjusting velocities above the ICB to fit new seismological observations. This fine-tuning decreased the IC radius estimate due to the direct trade-off between the thickness of the IC and OC and the P-wave velocity in both layers.

In order to understand the trade-off between layer thickness and layer velocity, we can imagine two layers: a liquid layer of P-wave velocity v_1 on top and a solid layer of P-wave velocity v_2 below. Let seismic waves enter from above and pass through both layers, travelling distance l_1 through the top layer and distance l_2 through the bottom layer. The total travel time through both layers is then $T = l_1/v_1 + l_2/v_2$. Now, if velocity v_1 is increased and velocity v_2 is kept constant, the total measured travel time will remain the same only if the second product becomes smaller, i.e. if l_2 decreases. Translated to the case of the IC, this means that the radius of the IC has to be readjusted to smaller values to counter the effects of seismic velocity increases.

For example, Bolt (1962) modelled a sharp decrease in velocity above the ICB instead as a gradual and constant velocity transition to produce the net effect of a higher P-wave velocity at the base of the OC. Therefore, the IC radius estimate had to become smaller (1216 km) to stay consistent with observed travel times. Buchbinder (1971) and Qamar (1973) obtained IC radii of 1226 and 1213 km, respectively. Engdahl et al. (1974) used PKiKP−PcP differential travel time data to estimate the radius of the Earth's IC to be 1220–1230 km while taking into account new estimates of the OC radius, though this was still in a direct trade-off with the P-wave velocity near the ICB. Studies in the 1980s concentrated on P-wave velocity structure in the uppermost IC; however; results varied widely. For example, Cummins and Johnson (1988) obtained significantly different estimates than the earlier work of Choy and Cormier (1983).

The existence of undulations in the ICB was speculated during the 1980s, although it was generally accepted from the observations of antipodal PKIKP waves and their travel times that there are no strong departures from spherical symmetry (Rial and Cormier, 1980), which was confirmed in the later study of Cormier and Choy (1986). However, Giardini et al. (1987) predicted undulations of more than 25 km based on an analysis of normal modes. This was at odds with the IC being in a state of hydrostatic equilibrium. Jeanloz and Wenk (1988) argued for convection in the IC, which might introduce relatively long-scale undulations along its boundary. Understanding the velocity and density contrast at the ICB was of growing interest to seismologists, for an accurate interpretation of the physical mechanism of IC growth would place valuable constraints on IC topography.

In a related study, Souriau and Souriau (1989) focused on the shape of the IC using new measurements of differential PKiKP and PcP travel times recorded at the Warramunga Array in the Australian Northern Territory (see Box 2.2 on the Warramunga seismic and infrasound array in Chapter 2). Their study added 11 new observations to the existing 14, and their measurements were made on recordings of earthquakes rather than nuclear explosions as in previous studies. PKiKP−PcP differential travel times at small epicentral distances are directly sensitive to the differential topography between the CMB and ICB (Figure 2.3c). Souriau and Souriau (1989) relied on the premise that if CMB topography is accounted for (from the model of Morelli and Dziewoński, 1987), topography of the ICB can be estimated. They found that the ICB is a near-perfect sphere that may exhibit a slight flattening. Their models exhibited a decrease of the polar radius between (1.6 ± 1.8) and (5.0 ± 1.5) with respect to a spherical shape and included a rotating IC. This view was in odds with the study of Giardini et al. (1987), who obtained significant ICB undulations of more than 25 km.

Parallel to the growing number of studies focused on the density contrast between the IC and OC, Poupinet and Kennett (2004) proposed an alternative solution to explain the coda of PKiKP waves. Instead of invoking IC scattering using high frequency PKiKP main arrivals (Vidale and Earle, 2000), they used a smaller, more consistent envelope. This secondary arrival did not build up over time as one would predict if it had scattered from within the IC, which implied the ICB was more complex than previously suggested, with possible lateral variations in both structure and topography. We will return to this finding in the next section.

As mentioned in Chapter 5, several studies of earthquake doublets argued for temporal changes in ICB topography as an alternative to IC differential rotation. According to this view, small-scale convection at the top of the IC causes localised topography at the ICB that is sensed by earthquake doublets sampling that same area of the ICB. Small differences in observed differential travel times would in this case be caused by rapidly changing topography at the ICB rather than changes in ray path relative to a rotating IC with a known velocity gradient. However, the presence of topography at the ICB does not rule out IC differential rotation. In fact, it is possible to use ICB topography as a marker to measure the rate of the rotation (Figure 3.5a). Using observations of SSI earthquake doublets recorded at the YKA array in Canada, Cao et al. (2007) argued for IC differential rotation and the presence of topography with a horizontal wavelength on the order of 10 km and a height of 300–500 m. Alternatively, in the absence of IC rotation, they invoked a time-changing topography (Figure 3.5b), in which case the viscosity of the IC would have to be relatively low ($<10^{16}$ Pa).

In 2012, Dai et al. (2012) argued for significant ICB topography. Based on observations of PKiKP−PcP travel time and amplitude variations, the authors modelled the ICB as having irregular topography with vertical variations up to 14 km within a lateral distance of less than 6 km. This disagrees with the geodynamical limit on dynamic ICB topography of a few hundred metres (e.g. Buffett, 1997; Monnereau et al., 2010; Alboussière et al., 2010; Deguen, 2012). Dai et al. (2012) argued that heterogeneous mantle structure near the CMB is unlikely to cause the observed anomalies, though they did not investigate other possible effects on travel times and amplitudes. In this class of seismological studies, it is often argued that the effect of the upper mantle and crust on differential travel times is minimal. However, Tkalčić et al. (2010) showed that crustal and upper mantle heterogeneity has a significant impact on the amplitude ratios of PcP and PKiKP waves and is an important source of amplitude fluctuations across arrays. Failure to account for these effects biases density estimates at the ICB.

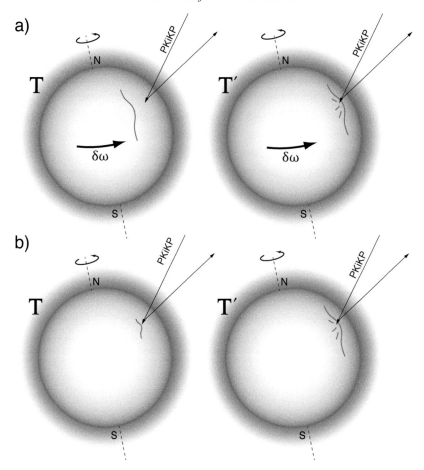

Figure 3.5 Schematic presentation of a method that uses repeating earthquakes and PKiKP waves for detection of ICB topography. Snapshots of IC at earlier time (T) and later time (T′) are shown on the left and right, respectively. The IC is shown by the central sphere embedded in the liquid OC. (a) The IC exhibits differential rotation $\delta\omega$ with respect to the mantle and ICB topography is static. (b) No differential rotation of the IC is assumed, but the ICB topography varies in time.

3.6 Frequency Dependent Reflection Coefficients and Constraints on ICB Topography

PKiKP waves are prominently expressed at high frequencies (Poupinet and Kennett, 2004; Tanaka and Tkalčić 2015). Poupinet and Kennett (2004) observed a number of PKiKP waves at the WRA array and other individual portable stations in Australia within the frequency band of 1–5 Hz. They concluded that energy channelling and low velocities near the ICB must be considered when explaining the character of the observed waveform codas. The associated mechanism called ICB scattering (ICBS) was different from that proposed by (Vidale and Earle, 2000), who concluded that the scattering occurs deeper within the IC. From the shape of the observed high frequency PKiKP waves, Poupinet and Kennett (2004) also concluded that the transmission of PKiKP through the D'' and CMB is efficient and without significant influence on PKiKP wavefield.

Tanaka and Tkalčić (2015) reported simultaneous observations of PcP and PKiKP waves from more than 10 events by the highly sensitive seismograph network in Japan (Hi-net), which consists of more than 700 stations. Visual inspection revealed that some PcP and PKiKP arrivals were visible at frequencies up to 5 Hz. Simultaneous observations of such a broad interval of high frequency content of PcP and PKiKP waves at a large number of stations is unprecedented and presents a new opportunity to increase constraints on the ICB density contrast.

Given the broad range of observed PKiKP frequencies, Tanaka and Tkalčić (2015) assessed complex frequency characteristics of the reflection coefficients rather than simply obtaining the amplitude ratio of PKiKP and PcP waves. Using Equation 3.4 they estimated reflection coefficients at the ICB from the corrected spectral ratios by considering reflection and transmission coefficients at the CMB, the geometrical spreading factor, and available elastic parameters. The spectral ratios were expressed as reflection coefficients as a function of frequency (Figure 3.6). The reflection coefficients peak around 2 Hz (group 8, 9, and 10) and 3 Hz (group 3), and exhibit minima around 1 Hz (group 2, 3, and 4) and 3 Hz (group 10). The spectra are divided into four general categories based on their frequency characteristics: (i) flat variation, where fluctuations in the relative strengths of the peaks and minima are between half and double the theoretical reflection coefficients (group 1, 5, 6, 7, and 11); (ii) a single distinct minimum (group 2 and 4); (iii) a single strong peak (group 8 and 9); and (iv) a strong peak and a minimum (group 3 and 10).

In a subsequent analysis, wave propagation above and below the ICB (in 2D) for frequencies up to 5 Hz was simulated via numerical calculation. Different modes of topography (e.g. sinusoidal and hyperbolic) with varying heights and scale lengths were tested through finite difference simulations. The grid spacing was 70 m to simulate high frequencies. The key assumption was that plane waves reflect from the ICB, so a sequence of point sources forming a line was used to simulate the wavefield and a series of virtual receivers above the ICB were used to record the reflected waves.

Results from simple tests precluded the existence of a liquid layer immediately above the solid IC. This suggested that the ICB is sharp and without an overlying thin melt layer due to a melting eastern hemisphere (see Section 3.11). Also precluded was topography of wavelengths and heights greater than a kilometre. This is compatible with geodynamical predictions of small, rather than extreme,

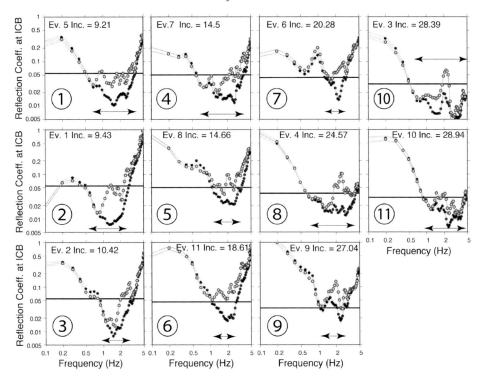

Figure 3.6 PKiKP-wave reflection coefficients at the ICB as a function of frequency, estimated from PKiKP/PcP spectral ratios (light grey circles) and noise/PcP spectral ratios (black circles). Ratios are corrected for fluctuations estimated from the reflection coefficients at the CMB obtained from the PcP/P ratios. Thin lines are upper and lower bounds of standard errors. Thick lines are theoretical values of the reflection coefficient calculated using model ak135 (Kennett et al., 1995). Horizontal arrows indicate the effective frequency ranges. Encircled numbers are groups of observations arranged by ascending incidence angle. Modified from Tanaka and Tkalčić (2015).

topography. However, more than one class of topography is required to explain the variety of observations. Lateral variations topography might signify lateral variations in IC solidification. If solidification is dynamically driven from top to bottom, its geographical pattern will be controlled by the pattern of OC convection (Bergman et al., 2002; Aubert et al., 2008; Gubbins et al., 2011). If the solidification is instead driven from the bottom up, the pattern will be affected by variations in IC convection (Deguen and Cardin, 2011). Furthermore, small-scale variations in topographic characteristics suggest small-scale convection in a mushy zone at the ICB (Bergman and Fearn, 1994; Deguen et al., 2007).

Considering entire waveforms, instead of just amplitude ratios, is an important step towards comprehensive analysis of the observed ICB-sensitive data. Further simulations exploring a wide range of geodynamical scenarios and more observations of PKiKP and PcP (especially in the western hemisphere) will improve our understanding of large-scale ICB structure and dynamics and provide critical insights about various open questions, such as that of the density contrast at the ICB, the extent and shape of ICB topography, and the origin of the scattering causing significant PKiKP codas.

3.7 State of the IC

Firm evidence for the solidity of the IC coincided with the first reports of PKiKP waves (reflecting from the IC), coming from a seismological study of Earth's free oscillations by Dziewoński and Gilbert (1971). They modelled eigenfrequencies of spheroidal modes with strong compressional and shear energy in the IC and observed theoretically predicted behaviour by either increasing IC rigidity or modelling the IC as liquid. An increase in rigidity would cause the bulk modulus to increase, in turn enhancing compressional energy while diminishing the shear energy of modes sensitive to the IC. An introduction of a liquid IC, however, implies a physically implausible, denser OC. Therefore, when comparing predictions and observations, the existence of a solid IC was favoured (Dziewoński and Gilbert, 1971).

PKJKP waves are elusive and continue to fascinate observational seismologists. Several studies reported new observations of these waves, but much of the seismological community remains sceptical. After Okal and Cansi (1998) opened a series of new studies on this subject, Deuss et al. (2000) argued that the previous observations of PKJKP waves were actually observations of simultaneous arrivals of pPKJKP and SKJKP waves. Cao et al. (2005) claimed new observations of PKJKP by the Graffenberg array and concluded that the S-wave velocity profile in the IC is 1.5 per cent faster than in the PREM model (Dziewoński and Anderson, 1981), meaning PKJKP waves would arrive about 9 seconds sooner than predicted. Wookey and Helffrich (2008) reported two new observations of IC shear waves from stacked Hi-net data. Vespagrams of these data revealed clear energy maxima associated with PKJKP and pPKJKP waves and a few late arrivals, likely related to anisotropy.

PKJKP waves are difficult to observe since polarisation differences make the conversion coefficient of P- to S-wave energy quite small. Furthermore, anisotropy can sometimes split the S waves, thereby spreading the energy out in time and reducing the amplitude, as energy is shared across two interfering quasi-S waves on any component of motion. Shearer et al. (2011) investigated possible causes behind the difficulties associated with observing PKJKP waves and demonstrated that routine observations are extremely unlikely at periods greater than 10 s. Therefore, they claimed that previously reported observations must have resulted from either focusing effects or lower than expected IC attenuation. If the observations at short periods are valid, the quality factor for shear waves must be much larger than predicted by PREM (Dziewoński and Anderson, 1981). Stacks of short-period seismograms and complete synthetics may further unveil this phase in the future.

Even without robust constraints on S wave velocity in the IC, a number of studies hypothesised the existence of various fluid-related structures in the IC. For example, according to Loper (1983), a liquid fraction exists atop the IC and the underlying solid fraction rapidly grows with depth, increasing one order of magnitude in only several hundred kilometres. A mushy (in the context of the melt-fraction content) zone extending tens of kilometres below the ICB has also been predicted (Cao and Romanowicz, 2004). Interestingly, most of these studies considered only 1D IC velocity structure, despite numerous studies reporting 3D heterogeneity.

Singh et al. (2000) invoked liquid inclusions to explain the low rigidity of the IC and the discrepancy between body waves and normal mode attenuation estimates (e.g. Widmer et al., 1991). Singh et al. (2000) speculate that a small amount of liquid inclusions could be present in the upper few hundred kilometres of the IC, though this requires that the liquid is not squeezed out of the IC (Sumita et al., 1996). Belonoshko et al. (2007) argue that grain boundaries and defects at high temperatures and pressures are a primary cause of seismic observations of a low shear modulus, i.e. low shear wave velocity (Cao et al., 2005).

Grain size also affects viscosity, which in turn affects the gravitational coupling of the IC and heterogeneity in the lowermost mantle (Buffett, 1997). Grain size can also determine the level of IC solidification and the amount of liquid inclusions; therefore, it is one of the key points of consideration in all geodynamical models.

Various scales of IC heterogeneity have been reported in a well-sampled portion of the upper IC, ranging from quasi-hemispheric (e.g. Tanaka and Hamaguchi, 1997; Niu and Wen, 2001; Yu et al., 2005), regional (e.g. Kaneshima, 1996; Stroujkova and Cormier, 2004; Krasnoshchekov et al., 2005) to fine-scale (e.g. Creager, 1997; Cormier and Li, 2002; Leyton and Koper, 2007a; Peng et al., 2008). The mushy zone hypothesis is well established (e.g. Cao and Romanowicz, 2004), with the estimated grain-size being on the order of 1–2 km (e.g. Vidale and Earle, 2000). Leyton and Koper (2007a) estimated the scale-length of volumetric heterogeneities in the UIC to be on the order of 1–10 km. Even smaller grain sizes of only a few hundred metres were suggested by Calvet and Margerin (2008), and Cormier and Li (2002) estimated scale-lengths of several hundred metres, increasing with depth up to 10s of km, to exist near the top of the IC. Both of these estimates come from interpreting Q in the IC as a scattering Q that broadens the PKIKP pulse by transferring scattered high frequency into the later coda. Deguen et al. (2007) suggest that uncertainties in conjunction with iron conditions at high temperature/pressure make it difficult to estimate the thickness of a mushy zone.

3.8 Isotropic P-Wave Velocity Distribution in the IC

3.8.1 Hemisphericity in Isotropic P-Wave Velocity in the IC

Chapter 4 discusses longitudinal variations in IC anisotropy and how there is a strong trade-off between anisotropic and, to a lesser degree, isotropic velocity variations, owing to the imperfect volumetric body wave coverage of the IC. The impact on isotropic velocity structure is more limited since the IC, at least in the upper several hundred kilometres, is reasonably well covered by PKIKP and PKiKP waves. Assuming that IC material is isotropic (see the discussion on the observed lack of seismic anisotropy in the top part of the IC in Chapter 4) allows inferences on the distribution of isotropic velocity to be made more readily.

The concept of a hemispherical dichotomy (spherical degree one) of IC properties dates back to the early 1990s of the twentieth century. It is interesting to follow chronologically how the pieces of the puzzle came together as better quality data became available. In the early 1990s Shearer and Toy (1991) used travel time data from the ISC catalogue to argue that PKP(BC)−PKIKP travel time residual patterns can be explained either by heterogeneity or anisotropy in the IC, each with about 1 per cent variation, which they described as aspherical symmetry. This was revisited by Creager (1992), who analysed short-period seismograms on a global scale. Since the residuals in his study were widely distributed geographically and interspersed with normal (as predicted) residuals, he discarded the possibility that aspherical structure is due to large-scale isotropic heterogeneity. Creager (1992) instead argued for the existence of axisymmetric anisotropy in the upper part of the IC with a fast axis quasi-parallel to Earth's spin axis.

Tanaka and Hamaguchi (1997), however, did not find evidence from broadband waveforms for anomalous residuals in the eastern latitudes. This meant that a simple model of IC anisotropy quasi-parallel with Earth's spin axis was invalid. Their measurements revealed a clear harmonic degree-one variation in isotropic compressional velocity, with the quasi-eastern hemisphere (qEH) of the IC (43° E to 177° E) being approximately 1 second faster than the quasi-western hemisphere (qWH). Since they had only a small number of north–south PKIKP and PKP(BC) wave ray paths and because their qWH dataset comprised only the South Sandwich Islands earthquakes recorded in northern latitudes, the authors concluded that if the observed travel time anomalies are indeed due to the existence of IC anisotropy, then the anisotropy cannot be simple. They had solid evidence against PKP(BC)−PKIKP differential travel time residuals stemming from the near core–mantle boundary structure and/or from a tilt in the symmetry axis of IC anisotropy. Because they did not have data against the hypothesis that cylindrical anisotropy existed only in the qWH, they accepted it. They proposed that the hemisphericity of isotropic compressional velocity reflects ancient core dynamics soon after planetary formation.

PKIKP waves whose travel times are measured differentially with respect to PKP(BC) travel times only sample the uppermost 400 km of the IC (UIC). Thus, evidence for a hemispherical dichotomy from the PKP(BC)−PKIKP differential travel time data only pertains to the UIC. It is difficult to probe deeper because the signal in PKP(AB)−PKIKP differential travel times cannot be easily decoupled from lowermost mantle structure, through which ray paths of PKIKP and PKP(AB) waves differ significantly. The hemispherical dichotomy was confirmed by studies using similar travel time data, though with augmented global coverage (e.g. Creager, 1999; Garcia and Souriau, 2000; Tkalčić et al., 2002). The top of the IC can also be studied with differential PKiKP and PKIKP travel times (e.g. Niu and Wen, 2001; Yu et al., 2005). Advantageously, the ray paths of PKiKP and PKIKP are nearly identical outside the IC; therefore, travel time residuals can be attributed to the IC. Unfortunately, differential PKiKP and PKIKP travel times only sample down to about 85 km beneath the ICB. Nonetheless, this independent dataset confirms the hemisphericity of the isotropic velocity structure.

3.8.2 Lateral Variations in Isotropic P-Wave Velocity in the IC

Various seismological studies have investigated IC quasi-hemisphericity (e.g. Tanaka and Hamaguchi, 1997; Creager, 1999; Garcia and Souriau, 2000; Niu and Wen, 2002; Irving and Deuss, 2011), radial variation in its structure (e.g. Ishii and Dziewoński, 2002; Calvet et al., 2006; Cormier and Stroujkova, 2005), and regional variation in its structure (e.g. Isse and Nakanishi, 2002; Helffrich et al., 2002; Tkalčić, 2010). Sharp radial and lateral velocity gradients in the UIC have been inferred beneath Central America (e.g. Creager, 1999) and the Indian Ocean (Stroujkova and Cormier, 2004). These studies led to a more complex view of heterogeneity distribution in the IC than was presented in Section 3.8.1.

A dichotomy in isotropic P-wave velocity and seismic attenuation between the two hemispheres of the IC and the conceptual framework built around this IC hemisphericity are largely based on travel-times and amplitudes of body waves sampling the IC in the equatorial region. More recently, Ohtaki et al. (2012) demonstrated that the IC does not have simple hemispherical variation. Their analysis was based on a unique sampling of the IC under Antarctica and evinces a single 'eyeball-shaped' anomaly in a small region beneath eastern Asia. According to Ohtaki et al. (2012), the fast qEH does not extend to the southern latitudes, and the region of high isotropic velocity is confined to an isolated region of equatorial east Asia. This contrasts the spherical degree-one distribution proposed by many authors. For a recent example, Waszek and Deuss (2011) calculated isotropic and anisotropic velocity models with a simple division between the qEH and qWH with a more complex layered structure resulting from differences in OC convection and

IC solidification rate. Tanaka (2012) determined that the hemispherical structure extends deeper than in previous models and has a depth dependency in the qE, and a constant Q in the qWH. Geballe et al. (2013) also argue for sharp hemisphere boundaries.

Cormier and Attanayake (2013) and Attanayake et al. (2014) collected PKIKP, PKiKP and PKP(BC) waveforms to study the UIC on a global scale, with special interest in the relationship between isotropic velocity and attenuation. Importantly, they isolated source radiation effects, which can have a significant impact on PKP waveforms. They found evidence for a more complex distribution of isotropic velocity in the UIC. According to Cormier and Attanayake (2013), there are two distinct areas of high velocity in the UIC: one located beneath the eastern equatorial Indian Ocean (stronger; deep extent), and another located beneath the Eastern Atlantic/West Africa region (weaker; shallow extent). Iritani et al. (2014) offer confirmation and moreover divide the qWH into two parts through the analysis of synthetic PKP waveforms. The new picture of lateral variations in P-wave velocity in the UIC thus contests hemispherical symmetry and hints at a more complicated UIC structure. Replacing spherical harmonic degree-one structure with spherical harmonic degree-two structure would strengthen the credibility of geodynamical models with thermal anomalies (harmonic degree two) from the lowermost mantle mapped onto the ICB and the structure of the UIC through OC convection (see Section 3.11).

As seismic data sampling the UIC accrues, the image of the IC is sharpened. But a piece of information that has been missing in all past studies that would significantly increase constraints on UIC structure is a PKP-wave observation by a set of recorders covering the entire epicentral distance range in which PKP waves sensitive to the IC can be recorded. This would not only provide unprecedented sampling of a given subvolume of the IC, but would also establish a means for modelling the entire empirical travel time curve, whose shape and slope are highly sensitive to the changes in velocity structure in the Earth's core. The establishment of Hi-net, a dense permanent array with regional-scale aperture, marked a new era in deep Earth seismology, and some of its pay-offs seen in a study of PKP waves that sample both quasi-hemispheres of the IC (Yee et al., 2014).

Up until the study by Yee et al. (2014), it was not possible to collect PKP travel time measurements from a single event on a sufficiently large number of stations across a range of epicentral distances as to allow data analysis similar to how travel-time data are used in reflection seismology. However, the high quality and geographical spread of approximately 800 Hi-net array recordings allowed Yee et al. (2014) to construct the complete empirical hodochrones (travel time curves) for differential travel times of PKP waves (PKP(BC)−PKIKP) near their triplication point for a large number of events. Four examples from the dataset of 47 events are shown in Figure 3.7. PKIKP and PKP(BC) arrivals are prominent on the records, and the empirical differential travel time curves can be compared with theoretical predictions from various models.

After other likely causes of the observed variation in PKP travel times were excluded (seismic anisotropy in the IC, lowermost and OC structures), the existence of strong regional variations in UIC P-wave velocity was suspected. Both PREM and ak135 were adequate reference models for lowermost OC velocity structure, and no significant modification of these models is necessary to explain the empirical PKP(BC)−PKIKP travel time curves. The most effective means of matching the observations was to modify P-wave velocity in the UIC along each ray path, as illustrated in Figure 3.8. The IC consequently appears to be laterally heterogeneous on a global (hemispherical pattern), regional (few hundred kilometres), and local scale (several tens of kilometres). The smallest of which is close to the estimated size of heterogeneity (on the order of 1 to 10 km: e.g. Leyton and Koper, 2007b) published in studies of PKiKP scattering in the upper IC. Yee et al. (2014) quantified the minimum and maximum magnitude of lateral perturbations of P-wave velocity with respect to ak135 and PREM to be 0.5–0.7 % (0.8–0.9 % with the lowermost mantle correction) in the qEH and 1.4–1.7 % (1.0–1.2 % with the lowermost mantle correction) in the qWH.

These results suggest that regional variations in isotropic P-wave velocity structure at the UIC are more widespread than previous studies were able to infer. At this point, it is difficult to determine if greater variation inferred for the qWH has any geodynamical backing since the sampling and spatial distribution is different for each hemisphere. Small-scale regional variations (from the observed variation in empirical PKP(BC)−PKIKP curves) may be superimposed on a large-scale hemispherical pattern in isotropic velocity or a spherical harmonic degree-two pattern. This scenario indicates the more complicated patterns of convection in the OC and possibly higher modes of convection in the IC (with processes fast enough to leave an imprint on IC growth) might play a more significant role than we anticipated.

Variations in convection and heat flux in the OC cause heat transfer variations across the ICB, which influence IC growth and crystal alignment (e.g. Yoshida et al., 1996). Regional variations in isotropic IC velocity are indicated by variations in crystal alignment (Creager, 1999), crystal texture (Stroujkova and Cormier, 2004; Cormier, 2007) and the random orientation of anisotropic patches (Calvet and Margerin, 2008). Complexity in the observed travel-time data for polar PKP ray paths (e.g. Leykam et al., 2010) suggests that the IC may be a conglomerate of anisotropic domains (Tkalčić, 2010; Mattesini et al., 2013).

Given the lack of concurrence amongst the results presented above, it appears that the characterisation and mapping of isotropic velocity heterogeneity in the IC will long remain an important, and unresolved, topic in seismology.

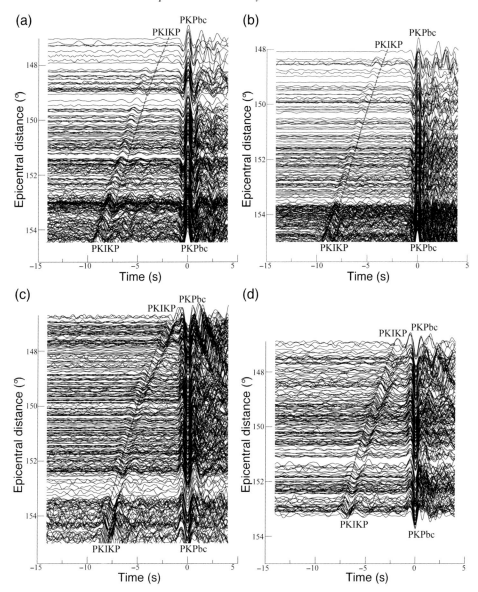

Figure 3.7 Selected teleseismic waveforms recorded by Hi-net stations aligned with respect to the detected PKP(BC) waves for (a) event 2008-07-01 (UT 01:54:40; $m_b = 5.6$; lat=$-58.251°$, lon=$-21.808°$, H=3.4 km) and (b) event 2008-10-04 (UT 12:11:18; $m_b = 5.7$; lat=$-59.452°$, lon=$-25.897°$, H=20.6 km) with ray sampling in the qEH and (c) event 2007-11-24 (UT 05:02:08; $m_b = 5.5$; lat=$-23.714°$, lon=$-68.821°$, H=96.3 km) and event 2005-07-26 (UT 14:11:34; $m_b = 5.7$; lat=$-15.339°$, lon=$-73.060°$, H=109.1 km) with ray sampling in the qWH of the IC. Dashed lines denote theoretical predictions of PKIKP-wave arrivals from model ak135 (Kennett et al., 1995). Modified from Yee et al. (2014).

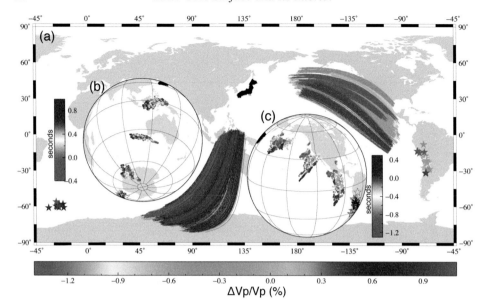

Figure 3.8 Surface projections of great-circle ray paths of PKIKP in the IC from the PKP travel time data used in Yee et al. (2014). Colours represent the best-fitting P-wave velocity perturbation from PREM (defining mod-PREM). Stars and triangles are the selected events and the Hi-net stations, respectively. (b) The entry, bottoming, and exit points of PKIKP in the IC are plotted in the qEH using the same colours as in (a). (c) Same as (b) but for the QWH. After Yee et al. (2014). (A black and white version of this figure will appear in some formats. For the colour version, please refer to the plate section.)

3.9 Seismic Attenuation in the IC

Seismic attenuation in the Earth can be viscoelastic (intrinsic) or a result of scattering. Scattering attenuation occurs when seismic waves reflect, refract, and convert at discontinuities in the Earth. Part of the seismic energy that travels from the source to the receiver is deflected or delayed by heterogeneities. PKIKP waves with frequencies around 1 Hz have wavelengths on the order of 10 km. The effects of scatterers less or equal to tens or hundreds of metres in size can be neglected. However, for any heterogeneity approaching or exceeding a kilometre in diameter, the effects will be significant. Further clarifying the distribution of heterogeneities within the IC is an important step in determining the significance of scattering attenuation in the IC.

Viscoelastic attenuation is energy loss due to internal friction and heat during the passage of earthquake waves. This can be described by a dimensionless factor (quality factor)

$$Q = \frac{W}{\Delta W},$$
(3.7)

where W is the average energy per cycle and ΔW is the energy lost or work done per cycle (Cormier, 2011). Small energy loss implies large Q and *vice versa*; thus it is intuitive to refer to the inverse of Q as attenuation.

Body wave attenuation is typically described using the attenuation parameter:

$$t^* = \int \frac{dt}{Q} = \sum_{i=1}^{N}, \tag{3.8}$$

where the integration is along the path in a layered Earth, dt is travel time segment (N segments in total), and t_i and Q_i are the travel time and quality factor for the ith layer (Lay and Wallace, 1995). The attenuation parameter measures the total travel time divided by the value of Q averaged along the path. Typically, for PKIKP waves the value of t^* is around $1s$.

We can rewrite Equation 3.7 by replacing energy with ground displacement amplitude A, as in

$$\frac{1}{Q} = -\frac{1}{\pi}\frac{\Delta A}{A}, \tag{3.9}$$

and from here it is possible to find a relationship between amplitude fluctuation and attenuation. Amplitude decreases by a fraction of $\frac{\pi}{Q}$ per cycle. In the frequency domain, the amplitude can be represented as

$$A(f) = S(f)\, G\, exp(-\pi f\, t_1^*), \tag{3.10}$$

where $S(f)$ is the source function, f is frequency, and G is a combination of effects such as geometrical spreading and transmission between the source and receiver (Niazi and Johnson, 1992). Denoting PKIKP and PKP(BC) with indices 1 and 2 and assuming that their ray paths differ only in the IC and that Q in the OC is infinity, it can be shown that

$$\frac{A_1(f)}{A_2(f)} \approx log\left(\frac{G_1}{G_2}\right) - \frac{\pi f t_{IC}}{Q_{IC}}, \tag{3.11}$$

where A_1 and A_2 are the amplitudes of PKIKP and PKP(BC) waves, and G_1 and G_2 are corresponding geometrical spreading and transmission effects associated with the PKIKP and PKP(BC) ray paths. t_{IC} is the total time the PKIKP wave spends in the IC, and Q_{IC} is the average attenuation in the IC. If Q_{IC} is not frequency dependent, the left-hand side of Equation 3.11 (the spectral ratio) is linearly dependent on the frequency, and Q_{IC} can be inferred from the slope of the logarithm of the amplitude ratios as a function of frequency (e.g. Niazi and Johnson, 1992; Souriau and Roudil, 1995).

The earliest studies of attenuation on global scale were conducted in the early 1900s (for a recent review, see Romanowicz and Mitchell, 2015). The motivation for IC attenuation studies originated from the discrepancy between the observed attenuation at high-frequency body waves and low-frequency normal modes, and from the need to estimate viscosity in the IC (e.g. Gans, 1972; Stevenson, 1981; Jeanloz and Wenk, 1988). The Q_μ^{-1} (shear attenuation) estimates from body wave observations (e.g. Doornbos, 1974) were relatively larger than those from spheroidal normal modes (Buland and Gilbert, 1978). Doornbos (1974) recognised the importance of considering frequency-dependent attenuation and argued that it increases as a function of depth by about 10 times from the ICB to the Earth's centre. The energy loss of radial modes $_0S_0$, $_0S_1$, $_0S_2$, etc. (a subset of spheroidal modes with no surface nodes and all motion in the radial direction) can be attributed exclusively to bulk modulus-related losses, as the vibrations are purely compressional. These majority of the energy of these modes is in the deep Earth, and measuring the bulk attenuation enables modelling of this parameter in the lower mantle and core (a highly non-unique problem). Estimates from the observations of PKJKP provide complementary constraints on shear attenuation Q_μ^{-1}. The early models of attenuation in the IC focused on viscoelastic attenuation, with two competing mechanisms: a shear mechanism from fluid flow due to partial melt inclusions, and a bulk mechanism representing a phase change induced during the passage of compressional waves, which would also require the presence of partial melt. Considering short period PKIKP waves, Cormier (1981) found that attenuation in the IC is consistently around 280, regardless of depth. A Q_μ^{-1} as high as 500–1000, which was found from PKJKP observations by Julian et al. (1972), would imply a lower Q_κ^{-1} (bulk attenuation), i.e. the attenuation mechanism in the IC would be based mainly on a phase change.

As PKIKP and PKP(BC) data were being collected and analysed, a global dataset of PKiKP and PKIKP waveforms was also analysed by waveform modelling to investigate the attenuation characteristics of the IC (Wen and Niu, 2002). Apart from confirming hemispherical differences in isotropic compressional velocity, the study revealed a hemispherical difference in attenuation. According to Wen and Niu (2002), the quality factor was larger in the qWH (600) and smaller in the qEH (250). A possible explanatory mechanism requires different geometric inclusions of melt and crystal alignment in the two hemispheres of the UIC. This model is supported by a study that showed how seismic velocity can depend on viscosity and on the fraction and geometry of the melt inclusions (Singh et al., 2000). Such a scenario would require a different intensity of UIC convection in each quasi-hemisphere, as driven by heat flow variation at the bottom of the OC. The viscosity would also have to be small enough for convection to develop within the solid IC. A limited radial extent of convection (confined to the UIC only) would preserve crystal alignment at greater depths of the IC, which was needed at the time to explain the widely accepted idea of cylindrical IC anisotropy. At the same time, convection confined within the UIC but with hemispherical differences would explain a well-documented observation of a hemispherically-varying thickness of the isotropic layer. This work sparked a new conceptual framework in which there is a spatial correlation between compressional wave velocity and the quality factor Q in the IC. A new image was emerging of an IC with a fast and more attenuating qEH and a slow and less attenuating qWH.

In the above study, the depth extent of the hemispherical variation in attenuation was not quantified due to the limited depth range of the sampling (the ICB to 85 km below the ICB). High attenuation in the qEH might extend deeper below the ICB than in the qWH if IC convection also drives a positive correlation between isotropic compressional velocity and quality factor Q (i.e. high velocity and high attenuation in the qEH and low velocity and low attenuation in the qWH). At the same time, Cormier and Li (2002) studied the radial variation of Q in the IC by analysing the forward scattering of pulse dispersion. The scattering Q broadens the PKIKP pulse by transferring scattered energy, characterised by high frequencies, into the later coda. Li and Cormier (2002) performed a variation on this analysis assuming viscoelastic attenuation. The idea of hemisphericity and a connection between isotropic velocity and attenuation had yet to develop at the time the aforementioned papers were published. Ignoring lateral variations, the papers reported depth-dependent attenuation in the IC, with a significant contribution (at least 25 per cent) from scattering rather than viscoelastic attenuation. This was an important piece of evidence that the distribution of grains and their varying size are an omnipresent phenomenon in the IC.

Cao and Romanowicz (2004) went a step further by estimating the attenuation quality factor on a global scale in the UIC, placing seismological observations within the context of rheological properties of the IC, such as porosity (melt fraction) and connectivity of liquid inclusions. Their analysis of PKiKP and PKIKP amplitude ratios in the time domain suggested the existence of a transition zone in the depth profile of Q in the qWH. According to their results, Q in the qWH decreases from near-infinite values at the ICB to ~210 at about 85 km depth beneath the ICB. It then increases with depth towards the centre of the IC. If a mushy zone is present in the UIC (Fearn et al., 1981), better connectivity of liquid inclusions (higher porosity) lowers compressional velocity and attenuation. They did not observe such a transition in the qEH, but they inferred that it must be located in the top 32 km of the IC (the upper limit of PKiKP and PKIKP ray path sampling in their study). The liquid inclusions are well isolated and the porosity is lower in the qEH, causing higher compressional wave velocity and higher attenuation than in qWH. They argued that the study of Stroujkova and Cormier (2004) supports this inference by finding a low velocity layer in the UIC in the qEH.

The observed hemispherical pattern and positive correlation between compressional wave velocity and attenuation were interpreted in light of the mechanism proposed by Sumita and Olson (1999), which predicts varying heat flow in the OC near the ICB. If higher porosity results from faster freezing (crystallisation) of the IC, the results of Cao and Romanowicz (2004) suggest that the qWH is colder and the qEH is warmer. This study also suggested that the signature of hemisphericity does not extend deeper than about 85 km beneath the ICB, which the authors interpreted as a direct constraint on the thickness of the mushy zone. Deguen et al. (2007) pointed out that uncertainties in conjunction with iron conditions at high temperature/pressure make it difficult to estimate the thickness of the mushy zone layer, but that it probably does not extend down to the centre of the IC, as there it would collapse under its own weight. They estimated that the size of the interdendritic spacing does not exceed a few metres.

Previous work revealed differences between shear attenuation estimated from body waves versus normal modes, the latter displaying lower values than the former (e.g. Masters and Gilbert, 1981; Fukao and Suda, 1989). Today it is generally accepted that any difference in IC attenuation estimates between body wave and normal mode studies must come from the frequency dependence of bulk attenuation (Andrews et al., 2006).

3.10 The Relationship Between Isotropic Velocity and Attenuation Structure in the IC

In two relatively recent studies of the isotropic velocity and attenuation in the IC, Cormier (2007) and Cormier et al. (2011) argue that outwardly stretched (radial) or parallel stretched (tangential) heterogeneity in the UIC reconciles certain observations of a hemisphericity in IC seismic attenuation and scattering of PKiKP-wave codas (Leyton and Koper, 2007b). Their results were in line with the hypothesis that the qEH is freezing faster than the qWH (Aubert et al., 2008), although they abandoned the textural stretching model in their most recent work, as will be described below. Although at odds with this scenario, a hypothesis of Monnereau et al. (2010) and Alboussière et al. (2010) that the qEH is melting faster than the qWH is consistent with the observed attenuation dichotomy between the two hemispheres of the IC (e.g. Cao and Romanowicz, 2004). Crystal size in the qEH might be larger than typical wavelengths of body waves sensitive to IC structure, thereby explaining the observed lack of strong back-scattering (Leyton and Koper, 2007b), while the melting component explains the higher attenuation. An alternative model could have an isotropic distribution of scale lengths where a PKiKP coda has been observed in the mid-Pacific, and possibly a larger crystal (or organised patch of crystals) in regions where no strong PKiKP coda has been observed (Cormier et al., 2011).

Most seismological studies that address the relationship between isotropic velocity and attenuation are confined to the upper portion of the IC. This is due to the fact that the differential travel time datasets with deeper PKIKP ray path sampling are more contaminated by structure outside the IC. This significantly obstructs further progress on understanding the relationship between isotropic and anisotropic velocity and attenuation in the deeper IC. Moreover, due to the uneven distribution of earthquakes and recording stations, and with most large earthquakes occurring in seismogenic zones located in moderate latitudes, the spatial sampling of polar regions of the IC is particularly poor (see Figure 2.7). However, in a recent study, Ohtaki et al. (2012) obtained a model of the IC beneath Antarctica that contains complexity that does not fit easily into the existing conceptual framework of hemispherical dichotomy. This is somewhat unsurprising given that the ray paths used in the study covered a region of the IC that had not been studied before. Their findings suggest a circular high compressional velocity anomaly confined to eastern Asia. In other words, the fast qEH does not extend to southern latitudes, and the region of high isotropic velocity is only an isolated region of equatorial east Asia. This more complicated lateral distribution of isotropic velocity in the UIC contrasts with the degree one distribution proposed by many authors. For example, in a relatively recent work, Waszek and Deuss (2011) calculated isotropic and anisotropic velocity models with a simple division of the qEH and qWH and a more complex layered structure resulting from differences in OC convection and the IC solidification rate. Tanaka (2012) determined that the hemisphericity extends deeper than in previous models, finding a depth dependency in the qEH, and almost a constant Q in the qWH.

Recently, Cormier and Attanayake (2013) and Attanayake et al. (2014) collected waveforms of PKIKP, PKiKP, and PKP(BC) seismic phases to study the UIC on a global scale, with a special interest in the relationship between isotropic velocity and attenuation. They found evidence for a more complex distribution of isotropic velocity in the UIC, different from a simple hemispherical dichotomy. According to Cormier and Attanayake (2013), there are two distinct areas of high velocity in the UIC, one located beneath the eastern equatorial Indian Ocean (stronger; deep extent), and another located beneath the Eastern Atlantic/West Africa region (weaker; shallow extent). These regions coincide with the regions of IC freezing predicted by Gubbins et al. (2011). According to Attanayake et al. (2014), there are three distinct regions of the UIC: 1) high velocity and high attenuation (qEH); 2) low velocity and low attenuation (eastern part of qWH); 3) low velocity and high attenuation in the central Pacific region (western part of qWH). In an independent study, Iritani et al. (2014) performed waveform analysis of PKP waves using synthetic seismograms and confirmed a division of the qWH into two parts. These new pictures of the UIC contradict hemispherical symmetry and hint at more complicated UIC structure. A departure from a harmonic degree one pattern in favour of a harmonic degree two pattern would substantiate geodynamical models where thermal anomalies (harmonic degree two) from the lowermost mantle are 'mapped' onto the ICB and UIC structure through OC convection.

3.11 Geodynamical Models Linking the IC with the Rest of the Earth

The spherical dichotomy of IC isotropic velocity and attenuation is a well-accepted, robust signal observed by a number of seismological studies (see Sections 3.8 and 3.9). This dichotomy presents a challenge to geodynamicists, and in recent years the deep Earth scientific community has witnessed the rise of several geodynamical models in designed to explain the observed hemispherical dichotomy at the top of the IC. Physical mechanisms behind the formation and maintenance of two discernible hemispheres that were in accordance with observed seismological were successfully identified. However, the models ranged between two completely opposite sides of the spectrum – some models predicted melting or slow growth to maintain the positive correlation between attenuation and velocity in the qEH (see Section 3.10) where other models predicted fast growth.

The carefully designed laboratory experiments (Sumita and Olson, 1999) and numerical simulations that followed these experiments (Aubert et al., 2008) showed that thermal heterogeneity in the lowermost mantle can explain asymmetric IC structure. The underlying argument was that the IC grows about $1\ mm\ yr^{-1}$ or, possibly, 2–3 times slower (Labrosse et al., 2001), meaning the top 100 km of the IC would solidify in about $100-300\ Myr$. During such a time interval, the most significant lowermost mantle structures (after having sunken from above during whole mantle circulation) would not have changed significantly (e.g. Torsvik et al., 2006). Thus, the cold regions in the lowermost mantle would draw heat out of the OC into the mantle at a higher rate than the warmer regions of the Pacific and Africa produce heat within the mantle. This, together with complex mixing in the OC, could create a hemispherically distinct UIC, which would act as a plausible interplay between the mantle and IC. However, figuring out how the observed pattern of UIC texture aligns with seismological observations remains an unresolved issue.

In the laboratory experiments of Sumita and Olson (1999), cold and warm fronts develop in the OC, which allows for the development of more rapid crystallisation on the cold qWH of the IC. Higher porosity was believed to produce material with a lower compressional velocity, and fast growth-induced anisotropy was argued to exist in the qWH (NB this interpretation of the relationship between the rate of growth and porosity and, in turn, the compressional wave velocity and related isotropy/anisotropy, is opposite from the interpretation given by Aubert et al., 2008).

Figure 3.9 illustrates a geodynamical model of Aubert et al. (2008), according to which the flow in the OC is dominated by upwelling and downwelling rotors around thermal anomalies. These convective columns connect the CMB with the ICB along the rotation axes. Consequently, the IC grows differentially, and the quasi-equatorial belt (slightly shifted northward from the equator) is characterised

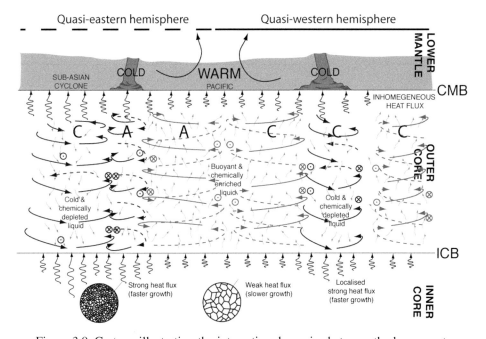

Figure 3.9 Cartoon illustrating the interaction dynamics between the lowermost mantle and IC and the resulting UIC dynamics according to Aubert et al. (2008). Lowermost mantle thermal heterogeneity creates a homogeneous heat flux and affects fluid flow in the OC, 'mapping' its lowermost mantle pattern onto the ICB. The connection between the lowermost mantle and the IC induces textural heterogeneity on the IC solidification front. Liquid flow in the OC is characterised by the presence of cyclones (with anticlockwise motion) and anticyclones, with the largest cyclone beneath the southeast Asia (sub-Asian cyclone). Downward motion (cold and chemically depleted liquid) is shown in blue, while upward motion (buoyant and chemically enriched liquid) is shown with red arrows. Dots and crosses correspond to the motion towards and away from the eye, respectively. Heat flux at the CMB and ICB is shown with curved arrows. Heat flux at the ICB is solely outward (from the IC to the OC). Texture of the IC corresponding to areas of fast and slow growth could be characterised by the existence of large and small grains. After Tkalčić (2015). (A black and white version of this figure will appear in some formats. For the colour version, please refer to the plate section.)

by faster growth than the rest of the IC. The area of fastest growth is located beneath SE Asia and corresponds to the sub-Asian cyclone that brings the cold fluid down to the ICB. Although this might create an impression of a harmonic degree-one distribution of the growth rate, their model is not strictly of harmonic degree one. For example, the northern parts of Africa and the central parts of the Atlantic Ocean and Americas are characterised by a faster growth rate, although half that beneath SE Asia. Whereas the thermal anomalies in the lowermost mantle invoked by tomographic studies of long wavelength shear waves are of spherical harmonic degree two, the lateral distribution of isotropic velocity and attenuation in the UIC at the time of Aubert et al. (2008) was claimed to be of a spherical harmonic degree one. Hence, the acceptable interpretation in the context of the observed IC spherical harmonic degree one dichotomy is that in the qEH, where the IC growth is faster, the solidification texturing occurs in the form of randomly spaced dendritic platelets.

According to metallurgic experiments with hexagonally packed iron alloys (e.g. Bergman, 2003), increased randomness in the orientation of platelets leads to a more isotropic medium, and higher compressional velocity and attenuation stem from the loss of energy by more pronounced scattering. To the contrary, a slower solidification of the IC results in wider spacing between platelets, which in turn gives rise to anisotropic material. More interdendritic fluid between platelets reduces compressional wave velocity, and attenuation will also decrease due to a smaller number of boundary crossings.

The model of Gubbins et al. (2011) is similar to that of Aubert et al. (2008), but it is driven by the possible existence of a variable-composition layer above the ICB (F-layer) (e.g. Poupinet et al., 1983; Cormier, 2009). In the geodynamical model of Aubert et al. (2008), the thermochemical wind in the OC creates lateral seismic anomalies above the ICB by inducing variations in its freezing rate, and therefore, the rate of light element release (compositional buoyancy), but this effect alone would not be sufficient to explain the existence of an F-layer. The effect produced in the model of Gubbins et al. (2011), however, is stronger. Figure 3.10 shows narrow downwellings and widespread upwellings. This situation creates both inward and outward heat flux from the OC to the IC, and results in melting of the IC. Melting releases heavy liquid and produces a variable-composition layer above the ICB. The existence of this layer does not require the IC to be locked to the mantle, although the simulations assume a locked geodynamo, with the narrow downwellings being locked at preferred latitudes. The models of Aubert et al. (2008) and Gubbins et al. (2011) agree that the qEH solidifies faster than the qWH. Gubbins et al. (2011) further interpret the IC fabric as having recently formed layers of unconsolidated mush on the freezing side (qEH).

In 2010 another mechanism was proposed to explain the hemispherical dichotomy observed in the UIC. Monnereau et al. (2010) and Alboussière et al.

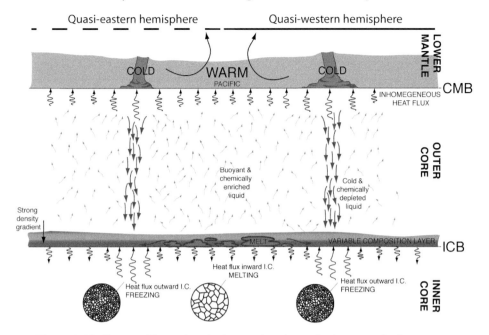

Figure 3.10 Cartoon illustrating the interaction dynamics between the lowermost mantle and IC and its consequence for the UIC and ICB according to Gubbins et al. (2011). Narrow downwellings (blue downward arrows) in the OC occur beneath cold regions of the lowermost mantle that coincide with the Circum-Pacific belt subductions. Wide upwellings (red upward arrows) occur beneath warm regions, which are associated with low shear-wave velocity in tomographic models of the lowermost mantle. Heat flux at the CMB and ICB is shown with curved arrows. Inward heat flux at the ICB results in melting of the IC, while the outward heat flux results in freezing. Texture of the IC corresponding to freezing and melting could be due to the existence of large and small grains, although they are not specifically invoked by the model. Heavy liquid released during melting forms the variable-composition layer immediately above the ICB. The heat flux at the ICB can be of both positive and negative sign, thus causing both freezing and melting. After Tkalčić (2015). (A black and white version of this figure will appear in some formats. For the colour version, please refer to the plate section.)

(2010) proposed that the IC is displaced from its centre of figure (geopotential centre) to keep its centre of mass co-located with the centre of Earth's mass. This model is supported by multi-scattering simulations that explain the positive correlation between compressional-wave velocity and attenuation in iron aggregates, similar to what has been observed for the IC. This requires small grains (about several hundred metres) for weaker attenuation and lower isotropic velocity, and large grains (typically about several kilometres) for stronger attenuation and higher isotropic velocity (Calvet and Margerin, 2008, 2012).

This model assumes that the ICB is a permeable boundary, meaning that the phase change permits material transfer. The regime for IC growth is always

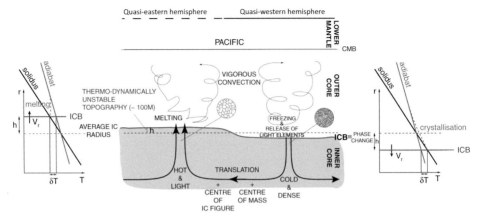

Figure 3.11 Cartoon illustrating the seismologically invoked hemispherical variation of the UIC according to Monnereau et al. (2010); Alboussière et al. (2010). According to this dynamical scenario, a superadiabatic IC with spherical harmonic degree one thermal heterogeneity (hemispherical dichotomy) undergoes displacement as to maintain equilibrium as its centre of mass shifts towards the colder, heavier side. Consequently, the ICB (brown solid line) develops positive topography h on the warmer side of the IC and negative topography $-h$ on the colder side of the IC. This topography is unstable since it is not at balance with the pressure and temperature conditions in the OC. Melting on the warmer side and crystallisation on the colder side counter the topography formation. Consequently, there is a permanent movement of IC material from the colder qWH to the warmer qEH. IC texture corresponding to areas of freezing and melting could be characterised by the existence of large and small grains such as in Figure 3.10. However, the main difference between this model and those described previously (see Figures 3.9 and 3.10) is that the melting occurs on the qEH, while the freezing occurs on the qWH. Diagrams on each side illustrate the thermodynamics at the ICB. The dashed red lines are average temperature profiles in the OC just above the ICB. The deviation from the adiabat stems from the latent heat. On the warmer side of the IC, the temperature just above the ICB is several milliKelvins lower, as required by the buoyancy forces that maintain the warmer hemisphere above the average IC radius (dashed black line). After Tkalčić (2015). (A black and white version of this figure will appear in some formats. For the colour version, please refer to the plate section.)

superadiabatic due to the difference in cooling rates between the faster cooling OC and the slower cooling IC, which promotes the onset of convection in the IC. The simplest mode of convection is a translational movement of IC material when a small thermal perturbation away from spherical harmonic degree one forms, as shown in Figure 3.11. Eastward translation induces positive topography on the eastern side and negative topography on the western side of the IC and means that material is gradually younger towards the east. The ICB topography (brown solid line) is both thermodynamically and gravitationally unstable. The hotter and lighter qEH experiences melting while the colder and denser qWH experiences crystallisation; these phase changes constantly act to remove the topography. The crystallisation on the colder side (qWH) constantly shifts the IC centre of mass towards that side. To combat the phase change, there is a viscous relaxation, which also acts to remove the topography.

Figure 3.11 features two diagrams, one on each side of the IC. On the melting side (left), the temperature at the ICB is lower than the adiabat and results in a cooling of the liquid above the ICB (red dashed line). A somewhat symmetric situation occurs on the opposite side, where the temperature at the ICB is higher than the adiabat (M. Monnereau, personal communication). The differences relative to the adiabat stem from the latent heat. The temperature difference at the ICB between the two sides of the IC is on the order of milliKelvins (Buffett, 1997). In fact, the warmer qEH has a slightly lower temperature than the colder qWH just above the ICB, which is required by the buoyancy forces maintaining the warmer hemisphere above the average ICB radius (shown by dashed black lines).

Alboussière et al. (2010) argued that IC translation and the consequent melting on one side and crystallisation on another produces a dense liquid iron depleted of light elements, which explains the seismologically observed low velocity gradient in the lowermost OC (Poupinet et al., 1983). IC translation can explain seismological observations of isotropic velocity and attenuation in the IC and also accounts for the existence of melting and a dense layer of liquid iron sitting on the top of the IC. Assuming this effect is real could help constrain viscosity in the IC, as is the topic of numerical work performed by Mizzon and Monnereau (2013). Their results are also interesting in the context of IC anisotropy. The translation of material inside the IC as proposed by Monnereau et al. (2010) requires IC viscosity to be above $10^{18} Pas$. However, a pure translation does not support the IC anisotropy invoked by numerous seismological studies (e.g. Deuss et al., 2010; Irving and Deuss, 2011). If the viscosity of the IC is lower than $10^{18} Pas$, higher modes dominate convection. For a viscosity lower than $10^{17} Pas$, translation is prevented by convective motions and a hemispherical dichotomy cannot develop. In the intermediate case hemispherical dichotomy would still develop and intense flow could possibly account for the observed asymmetry of IC anisotropy.

Seismologically observed ICB patterns play a significant role in experiments and simulations because they directly constrain, from the bottom side, OC flow and magnetic field structure. The dynamics of IC solidification is imprinted in its fabric, and seismological observations of that fabric are vital to the further advancement of knowledge of the deep Earth.

4

Inner Core Anisotropy

"The progress of science has always been the result of a close interplay between our concepts of the universe and our observations on nature. The former can only evolve out of the latter and yet the latter is also conditioned greatly by the former. Thus in our exploration of nature, the interplay between our concepts and our observations may sometimes lead to totally unexpected aspects among already familiar phenomena."

Tsung-Dao Lee, Nobel Lecture (Lee, 1957)

4.1 Introduction to Seismic Anisotropy

Seismic anisotropy[9] is a physical property of the Earth that describes the directional dependence of various seismic parameters, such as the velocity and attenuation of seismic waves. Today, it is generally accepted that some form of seismic anisotropy is present in the IC, in a similar way that it is in the Earth's crust and mantle, and in laboratory conditions under high temperatures and pressures, where it can be more directly examined.

Unsurprisingly, anisotropy is a widespread phenomenon encountered in nature and, therefore, in many fields of science and measurement techniques. For example, it occurs in the orientation of the magnetic field in plasmas, the growth of leaves, heat conduction, the random motion of water molecules in the brain, the formation of sediments, and the growth of crystals. We also encounter anisotropy across various branches of technology, such as spectroscopy, filtering, optics, sheet metal forming, and ultrasound imaging, among many other techniques.

The course of development of the IC anisotropy concept contains all elements normally present in a scientific process, where observations first lead to questions that then demand answers. The first and foremost question at the root of our

---

[9] *Note:* 'Anisotropy' originates from the late nineteenth century, from Greek *anisos* = '*unequal*' + *tropos* = '*turn*'.

fascination with IC anisotropy, "Is IC anisotropy the real mechanism behind seismological observations?", is a driving force for a number of seismological studies. Another relevant question, "If there is IC anisotropy measurable by existing seismological techniques and spatial coverage, how widespread and large is it?", naturally follows the first question. Consequently, many geoscientists wonder: "What is the physical mechanism behind IC anisotropy, and what can it tell us about the dynamics and history of our planet, how it formed, perhaps even how the geomagnetic field came to its existence and how it evolved?"

When the concept of IC anisotropy first developed in the 1980s, there was a notion that anisotropy in the velocity of the waves traversing the IC might be the most likely explanation for two independent classes of seismological observations: anomalous normal mode splitting and differential travel times. This was then followed by the establishment of IC anisotropy as a appealing conceptual framework. While seismologists looked for further evidence of IC anisotropy using seismic waveforms, geodynamo modellers and mineral physicists studied the conditions and mechanisms associated with anisotropy. Yet, as time went on, scientists began to realise the potential complexity hiding in the shadow of the current relatively simple image of IC anisotropy. As a result, doubts arose. Even the sheer existence of IC anisotropy was challenged as alternative hypotheses to explain seismological data came to surface. More complex models were proposed. New waveform data, new techniques, and new methods will provide increasingly accurate snapshots of IC anisotropy and answers to the questions posed here.

4.2 The Discovery

4.2.1 Evidence from Travel Times

In Chapter 2, we have seen that travel times of PKP waves were used in the construction of the spherically symmetric models of the Earth's interior, where seismic properties vary only with radius, not laterally. Spherically symmetric 1D reference Earth models (e.g. Jeffreys, 1926a; Dziewoński and Anderson, 1981; Kennett et al., 1995) do not consider lateral (isotropic or anisotropic) heterogeneity and treat the Earth's IC as a homogeneous sphere.

Ellipticity Corrections

Imagine that we wanted to use a new collection of carefully measured travel times of PKIKP waves to verify the accuracy of the published spherically symmetric models. We could do that by comparing the measured travel times with the predicted (theoretical) ones. In spherically symmetric models of the Earth's interior, the predicted travel times of PKIKP waves remain invariant for any two source-receiver points arbitrarily chosen on the globe as long as the epicentral distance

Table 4.1 *Earth's ellipticity corrections for PKIKP waves (given in seconds) for varying azimuth from source to receiver (A_{SR}) in columns, and epicentral distance (Δ) and source latitude (S_{lat}) in rows.*

	A_{SR}							
	0°	45°	90°	135°	180°	225°	270°	315°
$S_{lat} = 15°$								
$\Delta = 145°$	1.01	1.15	1.10	0.47	0.05	0.47	1.10	1.15
$\Delta = 160°$	1.21	1.22	1.09	0.75	0.56	0.75	1.09	1.22
$\Delta = 175°$	1.16	1.14	1.08	1.01	0.98	1.01	1.08	1.14
$S_{lat} = 45°$								
$\Delta = 145°$	0.30	0.17	−0.36	−1.19	−1.62	−1.19	−0.36	0.17
$\Delta = 160°$	−0.02	−0.15	−0.56	−1.08	−1.32	−1.08	−0.56	−0.15
$\Delta = 175°$	−0.50	−0.58	−0.67	−0.80	−0.85	−0.80	−0.67	−0.55
$S_{lat} = 75°$								
$\Delta = 145°$	−1.37	−1.49	−1.81	−2.17	−2.33	−2.17	−1.81	−1.49
$\Delta = 160°$	−1.90	−1.99	−2.21	−2.45	−2.55	−2.45	−2.21	−1.99
$\Delta = 175°$	−2.33	−2.36	−2.42	−2.48	−2.51	−2.48	−2.42	−2.36

and the depth of source are preserved. In reality, this is not the case because the Earth is not a perfect sphere. Therefore, the predicted travel times from spherically symmetric models need to be corrected for the Earth's ellipticity before we can compare them with our collection of measured PKIKP travel times. These corrections depend on the source latitude and depth, epicentral distance (angular distance between the source and receiver, where the coordinate system's origin is the Earth's centre), and azimuth (the angle measured in spherical coordinates on the Earth's surface from the source to the receiver). The variations associated with source depth can be neglected, and we will assume that the source is at the Earth's surface.

For the sake of demonstration, we can take an arbitrary grid of source latitudes and epicentral distances and vary the azimuth angle. The ellipticity corrections for PKIKP waves calculated with respect to the ak135 model (Kennett et al., 1995) are shown (in seconds) in Table 4.1. Although the grid is coarse, we can get a rough idea of required PKIKP travel time corrections. For example, for an epicentral distance of 160°, the appropriate ellipticity correction to be added to the PKIKP travel time predicted by ak135 model varies from –2.55 s for the source occurring in the polar region to 1.22 s for the source in the vicinity of the equator. This significant variation must be taken into account, as its magnitude is equivalent to the magnitudes of many phenomena that seismological studies attempt to address. Thus, only after the ellipticity correction is taken into account can one compare the measured with the predicted travel times.

Observations of Directional Dependence of Travel Times of PKIKP Waves

Let us now assume that a collection of globally measured PKiKP travel times empowered us to answer a simple question: "Does the IC have a constant radius?" Imagine that, even after ellipticity corrections are applied, our measured travel times of PKIKP waves significantly deviate from the predicted travel times. This could either be due to the fact that the real Earth is laterally heterogeneous, making our models oversimplified, or that our approximation of an Earth comprised of homogeneous shells is actually a good one, but the model's properties are not correctly determined. Let us assume we are dealing with the latter. We know that the spherically symmetric models are composed of laterally homogeneous layers with constant velocity and thickness. Thus, we have two possibilities: either the velocity of seismic waves in an arbitrary layer through which they pass is incorrect or the position of internal boundaries is inferred incorrectly. In the latter case, we can adjust the positions of boundaries in our initial Earth model until we achieve the best fit to the observed data. If this is done using a global collection of travel times of PKiKP waves that reflect of the ICB, it becomes clear that this method can be used like Sonar to probe the shape of the IC and decide whether the IC has a constant radius or significant variations in shape.

Early estimates of the IC radius (e.g. Engdahl et al., 1974) using the principle described in the previous paragraph did not reveal any peculiarities in the shape of the IC. However, Poupinet et al. (1983) noticed a significant excursion in differential travel times of PKIKP−P waves from predictions. The pattern of travel time residuals, when expanded in terms of spherical harmonics, exhibited a latitudinal dependence: polar stations recorded PKIKP waves travelling faster in comparison with equatorial-based stations.

Two examples of global observations of PKIKP waves are shown in Figures 4.1 and 4.2. The station in Tamarranset, Algeria, (TAM) is one of the highest quality stations worldwide (operated by the GEOSCOPE network), and has recorded a number of clear arrivals of PKIKP waves from the Tonga–Fiji region. These earthquakes occurred in a subduction zone environment, which has favourable radiation patterns for PKIKP waves. We can see that PKP phases were recorded at almost antipodal distances with unprecedented quality (Figure 4.1). What is special about this particular source-receiver geometry is that PKIKP waves traverse the IC almost perpendicularly to the Earth's rotation axis. The onsets of PKIKP waves are close to the predicted times by the spherically symmetric Earth model ak135 (Kennett et al., 1995), as indicated by the vertical dashed line. To contrast the observations at the TAM station, the earthquakes from the southern oceans recorded by the Siberian station Norilsk, Russia, (NRIL) are shown in Figure 4.2. These ray paths are nearly parallel to the rotation axis of the Earth. PKIKP waves are recorded at slightly

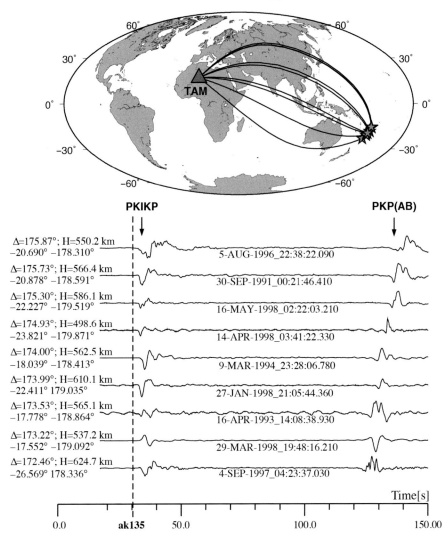

Figure 4.1 Location of selected earthquakes from the Fiji Islands region (stars) recorded at the station Tamanrasset (TAM), Algeria (triangle) and corresponding PKIKP ray paths projected on the surface of the Earth in map view. Seismograms (vertical components) are shown in their ascending order of epicentral distance from bottom to top. Arrivals of PKIKP and PKP(AB) waves are indicated by arrows (note that PKP(BC) is not seen in this epicentral distance range). Theoretical onsets of PKIKP waves from the ak135 model (Kennett et al., 1995) are shown by dashed vertical lines. Epicentral distance, hypocentre depth, origin time and location are listed for each seismogram. Modified from Tkalčić (2001).

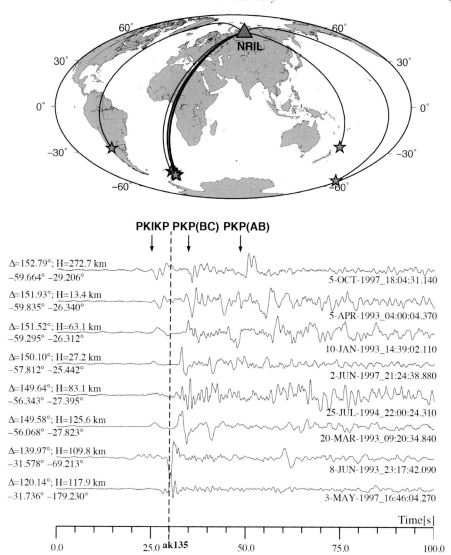

Figure 4.2 Location of selected earthquakes from the southern oceans (stars) recorded at the station Norilsk (NRIL), Russia (triangle) and corresponding PKIKP ray paths projected onto the surface of the Earth in map view. Seismograms (vertical components) are shown in their ascending order of epicentral distance from bottom to top. Arrivals of PKIKP, PKP(BC) and PKP(AB) are indicated by arrows. Theoretical onsets of PKIKP waves from the ak135 model (Kennett et al., 1995) are shown by dashed vertical lines. Epicentral distance, hypocentre depth, origin time and location are listed for each seismogram. Modified from Tkalčić (2001).

shorter epicentral distances than those at the station TAM, but their onsets are now notably advanced in comparison to ak135. These two examples, although showing the waveforms recorded during the 1990s, illustrate the nature of the observations made a decade earlier by Poupinet et al. (1983) quite well.

The observed pattern of travel time residuals in the study of Poupinet et al. (1983) resembled that of the geomagnetic field intensity. It became clear that a homogeneous sphere model of the IC cannot explain the measured travel time residuals. This led the authors to speculate that the IC might have a less dense material in the equatorial belt regions and denser material near the poles (Figure 4.3a). Alternatively, a prolate IC (a spheroid lengthened in the direction of a polar diameter) could create the same pattern of the measured travel times.

In the study described above, the authors made several assumptions; thus their intriguing results required further scrutiny. One of their assumptions was that the influence of heterogeneity present in the Earth's crust and mantle could be neglected since each differential travel time was derived from a pair of similar ray paths. This is a good approximation for the crust, as the ray paths of P and PKIKP waves are indeed nearly coincident. Therefore, the differential travel time is insensitive to the unknown heterogeneity in the crust, which influences the travel times of the P and PKIKP waves in the same way. Even a possible interaction with a subduction zone in the upper mantle might be neglected due to the close proximity of the P and PKIKP wave ray paths. However, these ray paths and their corresponding travel time integrals start differing significantly deeper into the mantle, especially in its lowermost part, which is heterogeneous on multiple spatial scales. Moreover, if there is significant topography at the CMB, it would influence travel times of PKIKP waves that refract at the boundary and those of P waves that bottom in the lowermost mantle.

Isotropic Heterogeneity Versus Anisotropy in the IC

Intrigued by the observation described in Section 4.2.1, Morelli et al. (1986) inverted the travel times of PKIKP waves reported in the International Seismological Centre bulletin to investigate isotropic heterogeneity patterns in the IC. When they corrected for whole mantle structure and CMB topography and assumed that all anomalous PKIKP travel times are caused by heterogeneity in the IC, they found a pronounced zonal pattern of heterogeneity for the waves sampling the deeper IC. This pattern was not present in the PKIKP data that sampled shallower parts of the IC. They concluded that it was physically implausible to explain 2–3 seconds of anomalous travel times with isotropic heterogeneity near the Earth's centre, which also disagreed with the splitting of the normal modes (Ritzwoller et al., 1986), and instead argued for a cylindrical pattern of anisotropy with a fast axis parallel to the rotation axis of the Earth while also allowing for isotropic heterogeneity in the

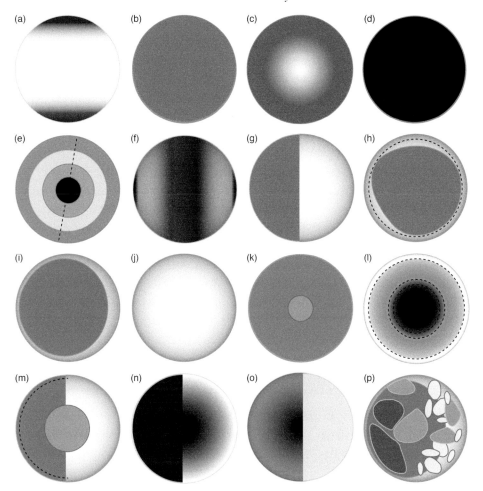

Figure 4.3 Selected conceptual models of IC anisotropy from published literature in chronological order. The spheres represent the IC viewed from the side,
with north on top and the western hemisphere at left. Shades are normalised in
each conceptual model so that darker shades represent stronger anisotropy, except
in (a), where darker tones at the poles represent denser material than in equatorial regions, and (p), where different tones represent different phases of iron.
Dashed circles represent transitions, and a dashed line in (e) marks a shift in the
fast axis of anisotropy relative to Earth's spin axis. The models conceptualise the
work of: (a) Poupinet et al. (1983), (b) Morelli et al. (1986), (c) Woodhouse et al.
(1986); Giardini et al. (1987), (d) Vinnik et al. (1994), (e) Su and Dziewonski
(1995), (f) Romanowicz et al. (1996), (g) Tanaka and Hamaguchi (1997), (h) Song
and Helmberger (1998), (i) Garcia and Souriau (2000), (j) Bréger et al. (1999,
2000); Romanowicz and Bréger (2000); Romanowicz et al. (2003), (k) Ishii and
Dziewoński (2002, 2003), (l) Cormier and Stroujkova (2005), (m) Sun and Song
(2008), (n) Irving and Deuss (2011), (o) Lythgoe et al. (2014), (p) Tkalčić (2010);
Mattesini et al. (2013).

IC. Obviously, a large trade-off would exist between the isotropic and anisotropic IC velocity structure when explaining PKIKP travel times, but the simplicity of cylindrical anisotropy was appealing, and from today's perspective, it remains a relatively uncomplicated IC model (Figure 4.3b). Please note that from hereafter the term *transverse isotropy* is interchangeable with *cylindrical anisotropy*, as the latter is a form of the former.

IC anisotropy was described by Morelli et al. (1986) in terms of cylindrical symmetry with respect to the Earth's rotation axis. It is necessary to parameterise the geometry of PKIKP waves traversing the IC in the co-ordinate system of this anisotropy. Therefore, a convenient way to map the geometry of PKIKP waves with cylindrical anisotropy is to express their sampling in the IC in terms of the angle that the PKIKP leg in the IC forms with the Earth's rotation axis (Figure 4.4).

Let us assume a PKIKP wave source at location $S(\theta_1, \phi_1)$ and a receiver at location $R(\theta_2, \phi_2)$, where θ_1 is the colatitude ($90° - latitude$) and ϕ_1 the longitude of the source, and θ_2 is the colatitude and ϕ_2 the longitude of the receiver. After some algebra (see Appendix B), it turns out that the cosine of the angle between the PKIKP leg in the IC and the Earth's rotation axis can be expressed as:

$$\cos\xi = \frac{\cos\theta_1 - \cos\theta_2}{\sqrt{2 - 2\cos\theta_1\cos\theta_2 - 2\sin\theta_1\sin\theta_2\cos(\phi_1 - \phi_2)}}. \tag{4.1}$$

In the above formula, ξ is defined within the interval $0° \leq \xi \leq 90°$. For the ray paths in the equatorial plane, $\xi = 90°$, while for the ray paths that traverse the Earth

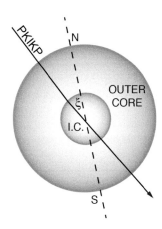

Figure 4.4 Geometry of PKIKP ray path in the IC. Earth's rotation axis is denoted by a dashed line pointing in the N–S direction. The angle between the PKIKP leg and the Earth's rotation axis is defined as ξ. For a detailed derivation of this angle, see Appendix B.

parallel to its spin axis, $\xi = 0°$ (note that we do not distinguish between the paths that have the source and receiver reversed, and thus form the same angle). $\xi \leq 30°$ is commonly referred to as being 'polar' and the corresponding PKIKP ray paths are called polar or quasi-polar paths. For a detailed derivation of this expression, see Appendix B.

For a 'weakly anisotropic' case, Morelli et al. (1986) considered a perturbation from an isotropic case. With some convenient substitutions for the elastic constants, it is possible to derive an expression for the velocity of PKIKP waves in a weakly anisotropic IC as a function of only two independent parameters, ϵ and σ.

$$v_p = \sqrt{\frac{A}{\rho}}(1 + \epsilon \cos^2\xi + \sigma \sin^2\xi \cos^2\xi), \tag{4.2}$$

where ρ is the density, $A = \lambda + 2\mu$, and λ and μ are Lamé constants (Kennett, 2001).

We refer the reader to Appendix C for a detailed derivation of the above expression and the assumptions made therein. Most researchers parameterise the velocity perturbation in a transversely isotropic medium in a slightly different form, i.e. by introducing a baseline shift γ_0, which represents uncertainties in Earth reference models. A purely isotropic case leads to $\rho v^2 = A$ (see Appendix C). Thus, Equation 4.2 can be written as

$$\frac{\delta v_p}{v_p} = \epsilon \cos^2\xi + \sigma \sin^2\xi \cos^2\xi + \gamma_0, \tag{4.3}$$

where $\frac{\delta v_p}{v_p}$ is a perturbation from a purely isotropic case with constant velocity v_p. Solving the above equation in ξ reveals that ϵ represents the velocity perturbation in the polar direction. Note, however, that if σ is extremely negative, the contribution of the second term in the above equation can become dominant, resulting in a minimum for intermediate angles ξ. In all published IC models so far that assume constant or layered cylindrical anisotropy, only the model of Ishii and Dziewoński (2003) has σ sufficiently large for the velocity perturbation to have a noticeable minimum (where the velocity perturbation in the slow direction of propagation is almost equal in amplitude to that for the polar, fast direction of propagation) at around $\xi = 45°$.

Alternatively, using simple trigonometrical rules, Equation 4.3 can be rewritten as

$$\frac{\delta v_p}{v_p} = \frac{1}{2}(\epsilon + \sigma)\cos^2\xi - \sigma \cos^4\xi + \gamma_0, \tag{4.4}$$

which is useful for the representation of velocity perturbations (or travel time perturbations) as a function of $cos^2\xi$. Another form of Equation 4.3 is

$$\frac{\delta v_p}{v_p} = \Phi P_2^0(cos\xi) + \Psi P_4^0(cos\xi), \tag{4.5}$$

where Φ and Ψ are linear combinations of the elastic constants and P_2^0 and P_4^0 are the associated Legendre functions. The above expression is particularly useful in that it illustrates that only zonal splitting function coefficients at degrees 2 and 4 are sensitive to transverse isotropy with a symmetry axis in the direction of the Earth's rotation axis. Ishii et al. (2002a) and Ishii and Dziewoński (2003) use the relationship to formulate a linear inverse problem. In other words, if travel time residuals are plotted on the globe with the angle ξ acting as colatitude (0°=pole and 90°=equator) and we assume that the entire travel time residuals signal is due to IC structure, these residuals should display a zonal pattern in order to remain compatible with the transverse isotropy hypothesis.

Calvet et al. (2006) derive a mathematical connection between parameters from various expressions for velocity perturbation, which is a useful tool when comparing various anisotropy trend and absolute magnitude calculations (see Table 4.2 near the end of this section).

4.2.2 Evidence from Normal Modes

The Earth as a Pipe Organ

Dziewoński and Gilbert (1971) confirmed that the IC is solid using observations of Earth's normal modes. In Chapter 2, we introduce normal modes and the inference of Earth structure through an analogy involving a pipe organ that makes noise when thrown down the stairs. If the inference of Earth's internal properties from normal modes sounds too difficult, compare it with deducing the deformation of the pipes within the organ after consecutive precipitations down a flight of stairs. This is exactly what was done during the 1980s.

To establish a good working definition of anomalous splitting of Earth's normal modes, let us assume that splitting is termed anomalous if the difference between the lowest and highest frequency singlet in a split multiplet is 50 per cent wider than predicted for a rotating Earth in hydrostatic equilibrium. As fitting examples, the predicted value for the mode $_{10}S_2$ is 6.0×10^{-6} Hz, while the measured splitting width for this mode is 14.0×10^{-6} Hz (Ritzwoller et al., 1986) (Figure 4.5), and the predicted value for the mode $_{18}S_4$ is 10.7×10^{-6} Hz, while the measured splitting width for this mode is 20×10^{-6} Hz (Widmer et al., 1992) (Figure 4.6).

As more data became available, a realisation occurred that some observations in the frequency domain, previously interpreted as peaks corresponding to separate

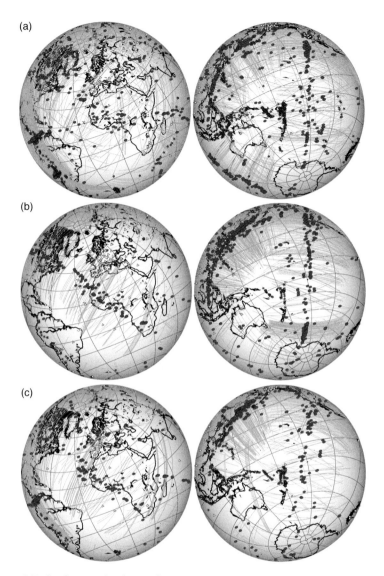

Figure 2.7 Surface projections of PKIKP ray paths through the IC are shown in orange on tilted perspective projections of the Earth centred on the western (left column) and the eastern (right column) hemisphere. Dark blue ellipses indicate the positions of the PKIKP ray bottoming points. Locations of earthquake sources are shown with red stars. Stations that recorded PKP(BC) and PKIKP waves used to compile the PKP(BC)−PKIKP differential travel time data are shown with yellow triangles. Data sets by (a) Tkalčić et al. (2002), (b) Garcia et al. (2006), and (c) Irving and Deuss (2011). (d) Souriau and Romanowicz (1997), (e) Garcia et al. (2006), restricted to shallow events only, and (f) combination of all data sets, with the exception of shallow events shown in (e). After Tkalčić (2015).

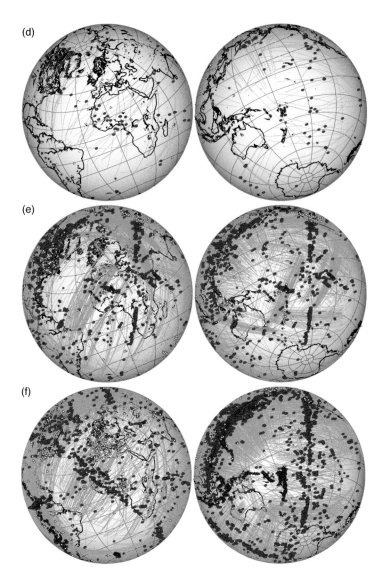

Figure 2.7 (*cont.*)

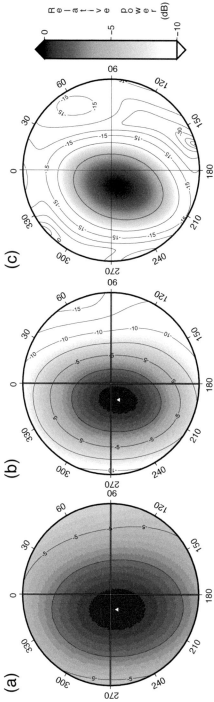

Figure 2.12 Observation of podal PKPPKP waves (Figure 2.3) and their precursors on the ILAR array for the event reported in Tkalčić et al. (2006). A strong signal is observed around slowness 0.22 $s/°$ and back azimuth $\theta = 244°$ for: (a) the main phase and (b) the precursor. (c) ARF for the slowness and back azimuth parameters observed in (a) and a monochromatic wave of 2 s.

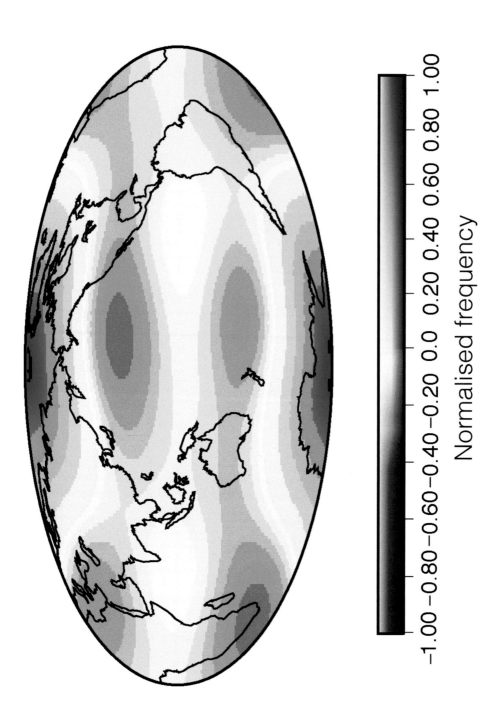

Normalised frequency

-1.00 -0.80 -0.60 -0.40 -0.20 0.0 0.20 0.40 0.60 0.80 1.00

Figure 2.20 Elastic splitting functions for the $_{13}S_2$ mode from the 1994 Bolivia and 1994 Fiji earthquakes shown as normalised frequency perturbations.

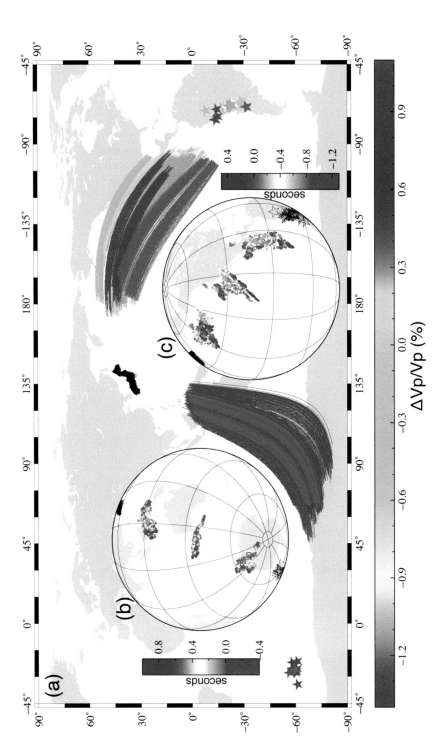

Figure 3.8 Surface projections of great-circle ray paths of PKIKP in the IC from the PKP travel time data used in Yee et al. (2014). Colours represent the best-fitting P-wave velocity perturbation from PREM (defining mod-PREM). Stars and triangles are the selected events and the Hi-net stations, respectively. (b) The entry, bottoming, and exit points of PKIKP in the IC are plotted in the qEH using the same colours as in (a). (c) Same as (b) but for the QWH. After Yee et al. (2014).

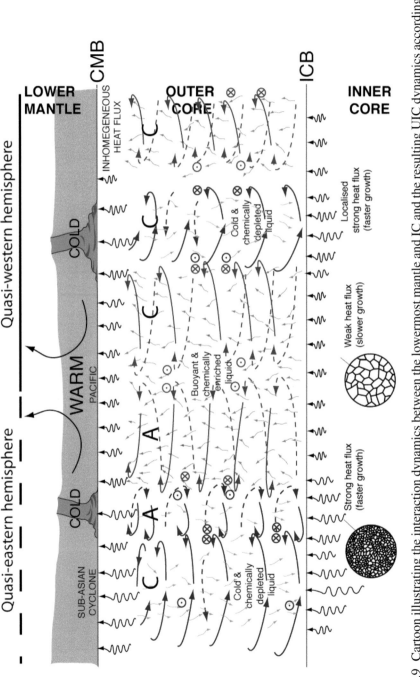

Figure 3.9 Cartoon illustrating the interaction dynamics between the lowermost mantle and IC and the resulting UIC dynamics according to Aubert et al. (2008). Lowermost mantle thermal heterogeneity creates a homogeneous heat flux and affects fluid flow in the OC, 'mapping' its lowermost mantle pattern onto the ICB. The connection between the lowermost mantle and the IC induces textural heterogeneity on the IC solidification front. Liquid flow in the OC is characterised by the presence of cyclones (with anticlockwise motion) and anticyclones, with the largest cyclone beneath the southeast Asia (sub-Asian cyclone). Downward motion (cold and chemically depleted liquid) is shown in blue, while upward motion (buoyant and chemically enriched liquid) is shown in red arrows. Dots and crosses correspond to the motion towards and away from the eye, respectively. Heat flux at the CMB and ICB is shown with curved arrows. Heat flux at the ICB is solely outward (from the IC to the OC). Texture of the IC corresponding to areas of fast and slow growth could be characterised by the existence of large and small grains. After Tkalčić (2015).

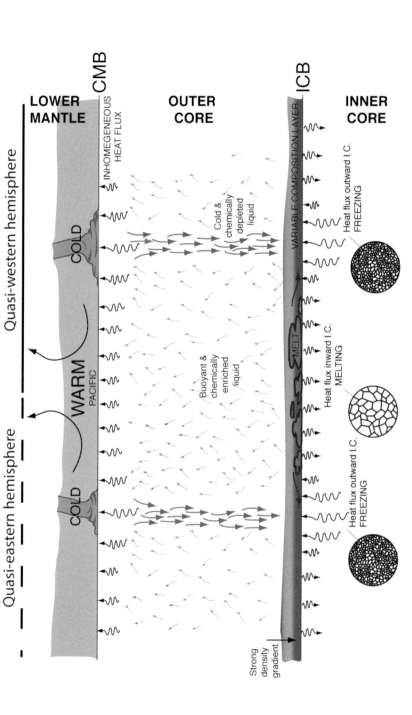

Figure 3.10 Cartoon illustrating the interaction dynamics between the lowermost mantle and IC and its consequence for the UIC and ICB according to Gubbins et al. (2011). Narrow downwellings (blue downward arrows) in the OC occur beneath cold regions of the lowermost mantle that coincide with the Circum-Pacific belt subductions. Wide upwellings (red upward arrows) occur beneath warm regions, which are associated with low shear-wave velocity in tomographic models of the lowermost mantle. Heat flux at the CMB and ICB is shown with curved arrows. Inward heat flux at the ICB results in melting of the IC, while the outward heat flux results in freezing. Texture of the IC corresponding to freezing and melting could be due to the existence of large and small grains, although they are not specifically invoked by the model. Heavy liquid released during melting forms the variable-composition layer immediately above the ICB. The heat flux at the ICB can be of both positive and negative sign, thus causing both freezing and melting. After Tkalčić (2015).

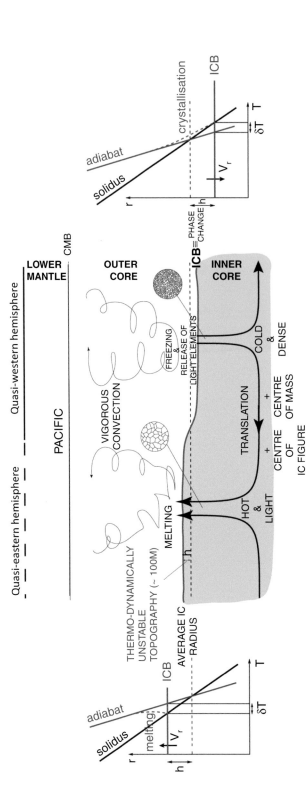

Figure 3.11 Cartoon illustrating the seismologically invoked hemispherical variation of the UIC according to Monnereau et al. (2010); Alboussière et al. (2010). According to this dynamical scenario, a superadiabatic IC with spherical harmonic degree one thermal heterogeneity (hemispherical dichotomy) undergoes displacement as to maintain equilibrium as its centre of mass shifts towards the colder, heavier side. Consequently, the ICB (brown solid line) develops positive topography h on the warmer side of the IC and negative topography $-h$ on the colder side of the IC. This topography is unstable since it is not at balance with the pressure and temperature conditions in the OC. Melting on the warmer side and crystallisation on the colder side counter the topography formation. Consequently, there is a permanent movement of IC material from the colder qWH to the warmer qEH. IC texture corresponding to areas of freezing and melting could be characterised by the existence of large and small grains such as in Figure 3.10. However, the main difference between this model and those described previously (see Figures 3.9 and 3.10) is that the melting occurs on the qEH, while the freezing occurs on the qWH. Diagrams on each side illustrate the thermodynamics at the ICB. The dashed red lines are average temperature profiles in the OC just above the ICB. The deviation from the adiabat stems from the latent heat. On the warmer side of the IC, the temperature just above the ICB is several milliKelvins lower, as required by the buoyancy forces that maintain the warmer hemisphere above the average IC radius (dashed black line). After Tkalčić (2015).

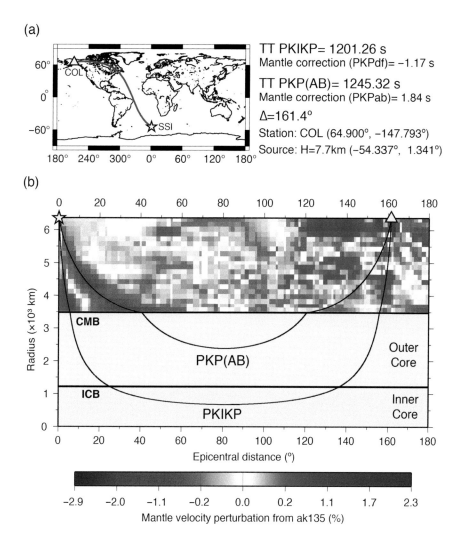

Figure 4.13 Illustration of the effect of mantle heterogeneity on PKP(AB)–PKIKP differential travel times. (a) Surface projection for a ray path from an earthquake in the South Sandwich Islands region to station College, Alaska (red line). Source and receiver information (star and triangle in the map), predicted total travel times through the Earth from ak135 model (Kennett et al., 1995) and corresponding corrections for mantle structure using the P-wave tomography model of Kárason and van der Hilst (2001) are given on the right. (b) Vertical cross-section through the Earth with major discontinuities (CMB: core-mantle boundary; ICB: inner core boundary) along the ray path shown in (a). Ray paths of PKP(AB) and PKIKP (black lines) through P-wave tomography model of Kárason and van der Hilst (2001). P-wave velocity perturbations from model ak135 are shown at the bottom. Modified from Tkalčić (2001).

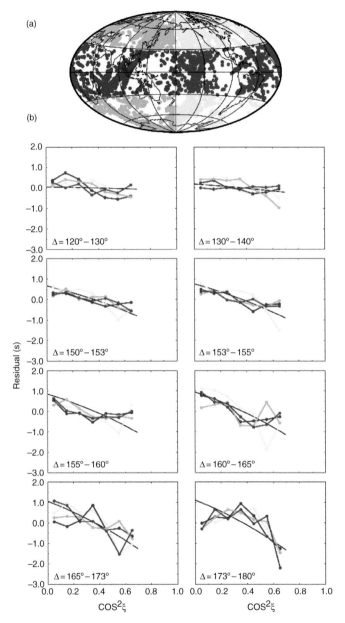

Figure 4.15 (a) Map of the geographic division of PKIKP data, designed to test the robustness of the IC transverse isotropy hypothesis. PKIKP data, based on their sampling of the IC, are divided into four regions: eastern polar (|latitude| > 30° ; 0° ≥ longitude ≥ 180°, indicated by green dots), eastern equatorial (|latitude| ≤ 30° ; 0° ≥ longitude ≥ 180°, blue dots), western polar (|latitude| > 30°; 180° ≥ longitude ≥ 360°, yellow dots), and western equatorial (|latitude| ≤ 30°; 180° ≥ longitude ≥ 360°, red dots). Dots on the map are bottoming points of PKIKP waves in the distance range 150°–153°. (b) Comparisons of subsets of PKIKP data at various distance ranges. Data with different colours correspond to quadrants shown in (a). Data are truncated at $cos^2\xi = 0.65$ because they are available for only two equatorial regions. The black curve is the prediction based on the constant anisotropy model from Ishii et al. (2002a). Modified from Ishii et al. (2002b).

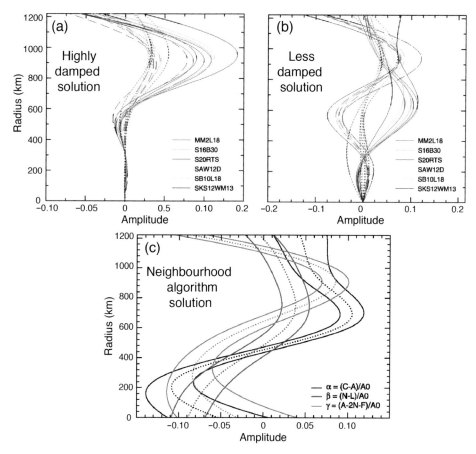

Figure 4.17 Models of anisotropy parameter strength as a function of radius in the IC (0 km = Earth's centre; 1221 km = ICB), resulting from linearised inversion with high (a) and small (b) damping. The solid lines represent P-wave anisotropy; the dotted lines, S-wave anisotropy; and the dashed lines, parameter γ. Different shear wave mantle modes were used to correct the data, yielding the inner-core models in coloured and grey areas. (c) Models resulting from the application of the Neighbourhood Algorithm (NA), where no damping is used. The thin dotted line represents the mean model, and thick surrounding lines correspond to two standard deviations taken from the posterior probability density functions obtained from the NA. Modified from Beghein and Trampert (2003).

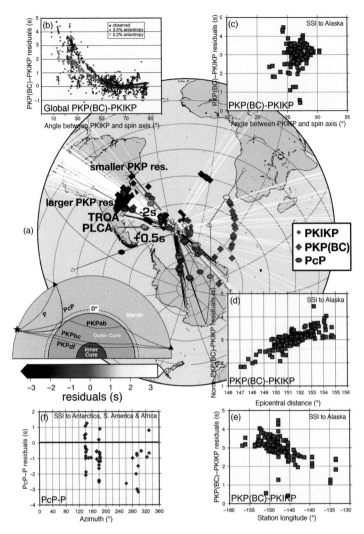

Figure 4.19 (a) Map of the location of SSI earthquakes used in studies of PKP(BC)−PKIKP and PcP−P differential travel times (stars). Reflection points of PcP waves at the CMB are projected to the surface (ellipses) in different colours corresponding to the observed PcP−P differential travel time residuals. Piercing points of PKIKP and PKP(BC) waves at the CMB are projected to the surface (small and large diamonds) with the corresponding PKIKP−PKP(BC) differential travel time residuals using the same colour scheme. Travel time residuals are relative to the model ak135 (Kennett et al., 1995). PKP and PcP ray paths projected to the surface are shown in white and black lines. Yellow lines indicate a corridor in which some of the largest departure from theoretical predictions in PKIKP−PKP(BC) and PcP-P travel times are observed. GSN stations TRQA (inside the corridor) and PLCA (outside the corridor) are highlighted. A schematic representation of Earth's cross-section and ray paths of seismic phases PKP, PcP, and P waves is shown in inset. (b) Global dataset of PKP(BC)−PKIKP differential travel time residuals as a function of the angle ξ along with the theoretical predictions from 2.2 per cent (yellow inverted triangles) and 3.5 per cent (red diamonds) models of uniform cylindrical anisotropy in the IC. PKP(BC)−PKIKP differential travel time residuals from SSI earthquakes as a function of: (c) the angle ξ, (d) azimuth, and (e) epicentral distance. (f) PcP−P differential travel time residuals from SSI earthquakes as a function of azimuth. Modified from Tkalčić (2010).

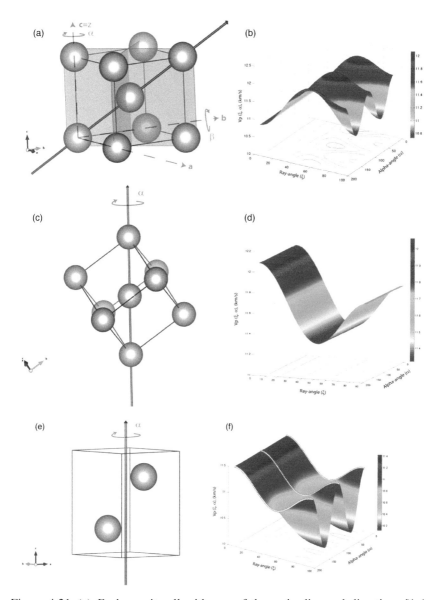

Figure 4.21 (a) *Fe–bcc* unit cell with one of the main diagonal directions [1 1 1] (red vector) representing the fast velocity axis. The two fast velocity planes [1 0 -1] and [1 1 0] at 90.0° with respect to each other are also shown in brown. The α and β angles are clockwise rotations about the *c*- and *b*-axis, respectively. (b) The periodicity of P-wave velocity (v_p) produced by a clockwise rotation about the Earth's spin axis (i.e. from $\alpha = 0°$ to $\alpha = 180°$) for the *Fe–bcc* model, with the fast velocity axis at 54.74° from the main *z*-axis ($\alpha = \beta = 0°$). (c) *Fe–$\overline{bcc}$* unit cell along the vertical crystallographic *c*-axis, aligned parallel to the Earth's rotational *z*-axis. (d) *Fe–$\overline{bcc}$* model, with the fast velocity axis (i.e. the main diagonal) oriented along the vertical crystallographic *c*-axis direction ($\alpha = 45°$ and $\beta = cos^{-1}1/\sqrt{3}°$). Note that there is no velocity modulation for a clockwise rotation about the *c*-axis, thus indicating that the system is purely transversely isotropic. (e) *Fe–hcp* unit cell and its [0 0 1] fast velocity axis direction. (f) v_p of *Fe–hcp* with the fast velocity direction along the vertical Earth's spinning axis ($\alpha = \beta = 0°$) showing the 90° periodicity produced by a clockwise *z*-axis rotation (i.e. from $\alpha = 0°$ to $\alpha = 180°$). After Mattesini et al. (2013).

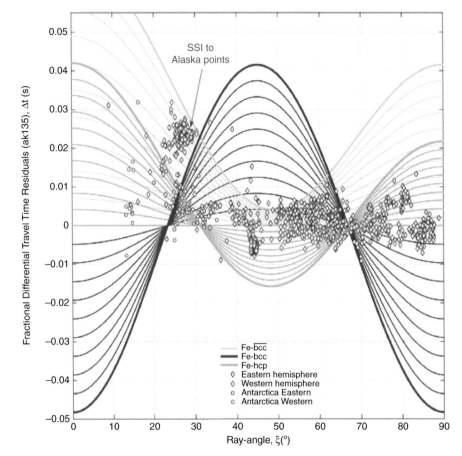

Figure 4.22 'Candy Wrapper' velocity model for PKP(BC)−PKIKP data points for a sampling depth range of 125.8–345.1 km. Data points are shown by blue symbols for the bottoming points in the qEH and by red symbols for the bottoming points in the qWH. The model consists of three sub-models: the conglomerate cubic phase (grey), the bare cubic phase (blue), and the hexagonal close packed iron (green). Note that when omitting the data from SSI region, there is a drastic strength reduction in elastic anisotropy from 3 per cent to 0.7 per cent. After Mattesini et al. (2013).

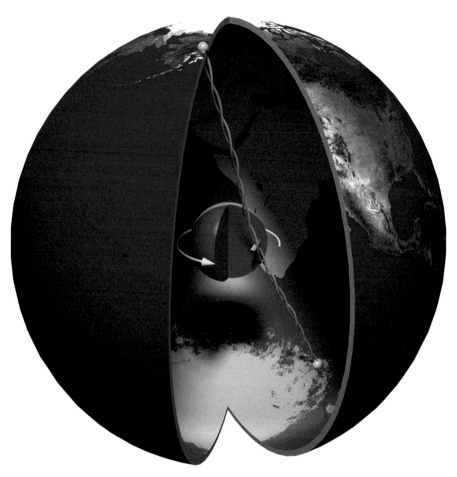

Figure 5.3 Illustration of the rotational dynamics of the IC. Marked are the South Sandwich Islands archipelago (the location of earthquake doublets) (yellow ball) and the COLA seismological station near Fairbanks, Alaska (green ball). The IC is shown at the Earth's centre. The PKIKP ray paths for an earthquake doublet are shown in blue and red. They traverse the western hemisphere of the Earth's IC (for a definition of hemispheres, see Section 3.8.1) through a well-studied portion featuring a linear gradient in isotropic velocity, which is shown by a small patch of varying colour. As the IC rotates eastwardly with respect to the mantle (indicated by the yellow arrow), a fixed source-receiver path will sample the same mantle and OC (not shown) structure, but slightly different IC structure due to the velocity gradient. If velocity gradient is used as a 'marker', it is possible to estimate the differential rate of the IC rotation from the measured time differences between the onsets of PKIKP and PKP(BC) waves (see the text).

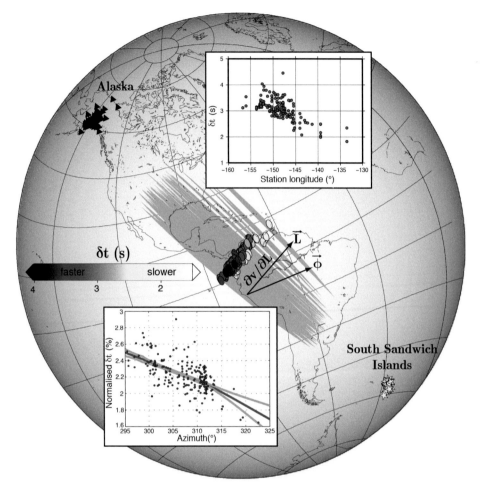

Figure 5.4 The location of the South Sandwich Islands earthquakes (yellow stars), stations in Alaska (triangles), and bottoming points of PKIKP waves (ellipses) are projected to the surface. Bottoming points of PKIKP waves are shown in colours corresponding to the travel time residuals, δt, defined in Chapter 2 in the box on PKP travel time residuals (Box 2.1, Equation (3)). Great circle path projections of PKIKP waves in the IC are shown by grey traces. The direction of the steepest gradient in velocity, $\frac{\partial v}{\partial L}$, is marked by $\vec{L}$, while $\vec{\phi}$ marks the changing longitude. The diagram at the top presents the time residuals, δt, as a function of the recording station longitude. The residuals normalised by the time the PKIKP ray spends in the IC are shown as a function of azimuth (measured from the source to the station) in the diagram at the bottom. The expected regression model obtained by averaging all resulting models in a transdimensional Bayesian inversion (see Appendix D) is displayed by the red line. Grey lines represent one standard deviation of the ensemble at each azimuth. Adopted from Tkalčić et al. (2013a).

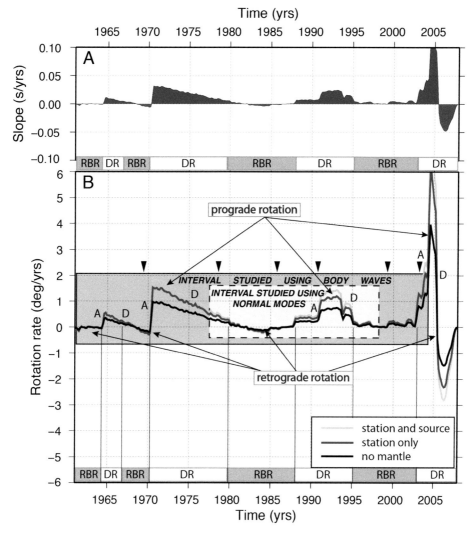

Figure 5.14 (a) The slope parameter (for explanation, see Equation 5.5) as determined by Bayesian inversion, in which the total time interval is divided into a variable number of constant (one-dimensional) Voronoi partitions separated by sharp discontinuities. The unknown parameters in the inversion are the number of Voronoi nuclei, their position along the time axis, and the regression value within each cell. Blue areas indicate eastward rotation whereas red areas indicate westward rotation of the IC with respect to the mantle. (b) Differential rotation rate of the IC (Equation 5.5) as derived from the slope for three different values of the velocity gradient (depending on the mantle model used for corrections) defined in Song (2000a). The dark grey and light grey rectangles delineate the time intervals used in the study of doublets (Zhang et al., 2005) and normal modes (Laske and Masters, 2003). The areas under these curves were integrated to estimate the total shift. The position of the reported geomagnetic jerks is indicated by black inverted triangles. 'DR' stands for 'differential rotation' and 'RBR' stands for 'rigid body rotation'.

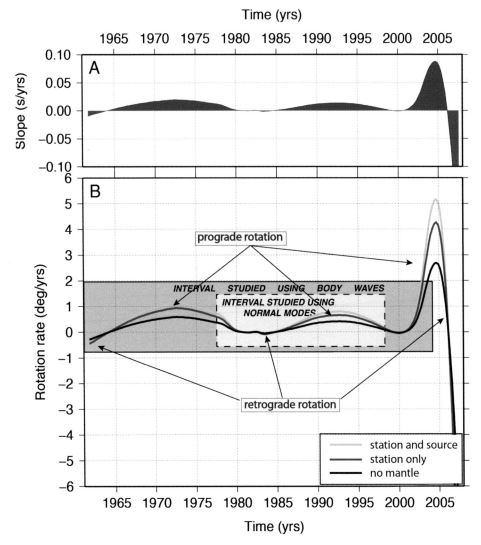

Figure 5.16 (a) The slope parameter (for explanation, see Equation 5.5) as determined by the Bayesian inversion of Tkalčić et al. (2013b), where the time interval is divided into a variable number of B-splines, which itself is a free parameter in the inversion. Blue areas indicate eastward rotation whereas red areas indicate westward rotation of the IC with respect to the mantle. (b) Differential rotation rate of the IC (Equation 5.5) as derived from the slope for three different values for the velocity gradient (depending on the mantle model used for corrections) defined in Song (2000a). The dark grey and the light grey rectangles delineate the time intervals used in the study of doublets (Zhang et al., 2005) and normal modes (Laske and Masters, 2003). The areas under these curves were integrated to estimate the total shift. Adopted from Tkalčić et al. (2013b).

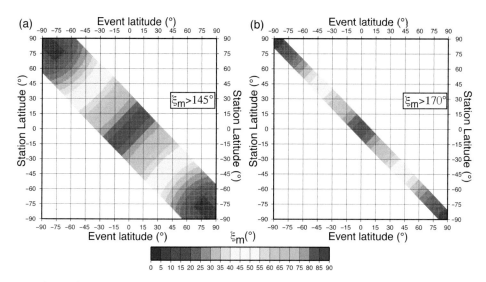

Figure 6.1 (a) Map of the smallest possible angle between PKIKP waves in the IC and the rotation axis of the Earth (ξ) for any given source-receiver pair at the Earth's surface under the restriction that $\Delta \geq 145°$, where Δ is epicentral distance. Source and receiver latitudes are plotted on the horizontal and vertical axes. Colours correspond to different values of ξ. White areas are source-receiver pairs for which the geometry of the PKIKP waves does not satisfy the above condition. (b) Same as (a) but under the restriction that $\Delta \geq 170°$. After Tkalčić (2015).

Figure 6.2 3D illustration of PKIKP ray paths traversing the IC from a view-point along the equatorial plane within the Earth. The yellow-orange globe in the centre of each image is the IC, and the map of the Earth's surface is projected onto the IC for orientation purposes. The ray paths are from existing datasets of PKP(BC)−PKIKP differential travel times collected through waveform correlation (Tkalčić et al., 2002; Leykam et al., 2010). The colours of the ray paths correspond to different values of travel time residuals: blue marks fast, white marks neutral, and red marks slow paths through the IC. Orange and yellow colours represent quasi-western and quasi-eastern hemispheres (qWH and qEH) of the IC, as defined by Tanaka and Hamaguchi (1997) (see Section 4.3.3). The IC as centred on the qEH and sampled by (a) quasi-polar PKIKP ray paths, defined by angle $\xi \leq 35°$ and (b) quasi-equatorial PKIKP ray paths, defined by angle $\xi \geq 35°$. The IC as centred on a transition between the qWH and qEH, sampled by (c) quasi-polar PKIKP ray paths, and (d) quasi-equatorial PKIKP ray paths.

Figure 6.3 3D illustration of PKIKP ray paths traversing the IC similar to Figure 6.2, but the IC is now viewed from a perspective within the Earth along the north-south axis. The IC as viewed from a point beneath the north pole, sampled by: (a) quasi-polar PKIKP ray paths, defined by an angle $\xi \leq 35°$ and (b) quasi-equatorial PKIKP ray paths, defined by an angle $\xi \geq 35°$. The IC as viewed from a point beneath the south pole, sampled by: (c) quasi-polar PKIKP ray paths, and (d) quasi-equatorial PKIKP ray paths.

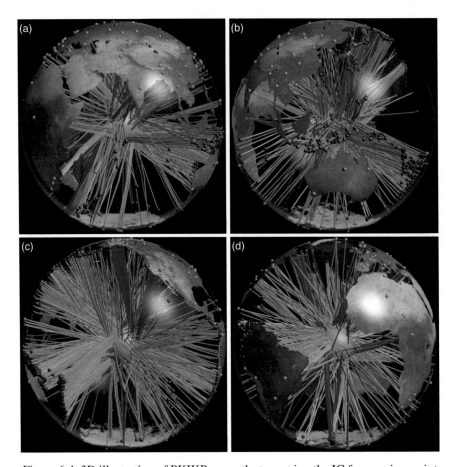

Figure 6.4 3D illustration of PKIKP ray paths traversing the IC from a view point outside the Earth in the equatorial plane. The yellow globe in the centre of each image is the IC. The ray paths are from the existing datasets of PKP(BC)−PKIKP differential travel times collected through waveform correlation (Tkalčić et al., 2002; Leykam et al., 2010). Colours of ray paths correspond to different values of travel time residuals: blue marks fast, white marks neutral, and red marks slow paths through the IC. Green balls are station locations and red balls are event locations. The view is centred on (a) Indian Ocean, (b) southeast Asia, (c) Pacific Ocean and (d) Atlantic Ocean.

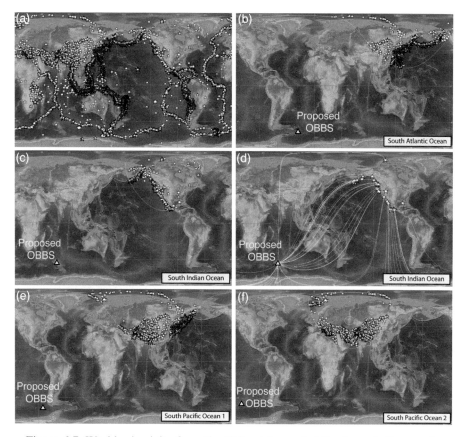

Figure 6.7 World seismicity from the ISC catalogue (white stars: shallow earthquakes, yellow stars: intermediate earthquakes, and red stars: deep earthquakes). (a) global earthquakes available in the ISC catalogue for the period 2000–2010. (b) a hypothetical station location in the South Atlantic Ocean near the South Sandwich Islands and all earthquakes in the epicentral distance range $145° \leq \xi \leq 175°$ that can potentially yield PKP observations. (c) the same as (b), but the hypothetical station location is in the South Indian Ocean near the Kerguelen Islands. (d) the same as (c), but only with earthquakes with $m_b \leq 5.5$ and including the corresponding great circle paths. (e) the same as (b), but the hypothetical station location is in the eastern part of South Pacific. (f) the same as (b), but the hypothetical station location is in the central part of South Pacific. Modified from Tkalčić (2015).

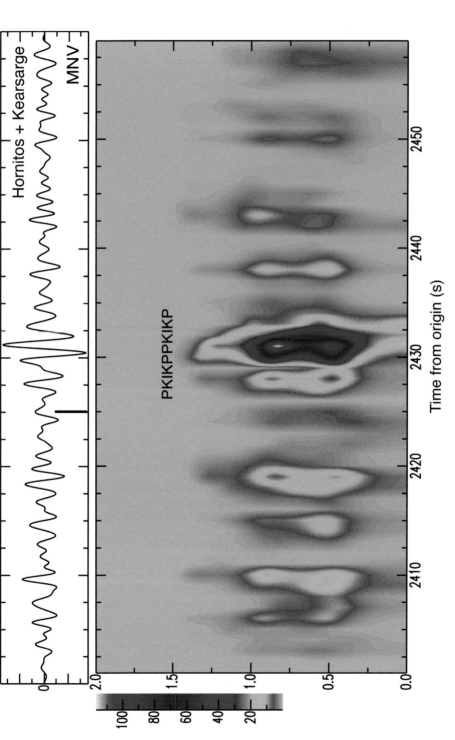

Figure 6.9 An observation of PKIKPPKIKP waves at station MINA, Nevada, after a zero-phase stack of the waveforms produced by two 150 kT nuclear explosions (Kearsarge, 17 August 1988 and Hornitos, 31 October 1989) (Tkalčić and Flanagan, 2004). The average epicentral distance from the explosions to the station is slightly less than 2°. A clear onset of PKIKPPKIKP is shown both in the time domain (seismogram on the top) and in the frequency domain (spectrogram at the bottom). The thick vertical line at 2425 s is the predicted travel time from ak135 (Kennett et al., 1995). The ray path associated with this observation samples the Earth about 50 km from its centre.

Table 4.2 *Summary of various studies assuming cylindrical anisotropy in the IC with some model characteristics.*

Source	Method/data type	Character of cylindrical anisotropy	Strength[a] of cylindrical anisotropy $\widehat{V_p}[\%]$
Morelli et al. (1986)	abs. travel times PKIKP	uniform	3.5
Woodhouse et al. (1986)	normal mode splitting	uniform radially varying	3.35 5.2 (ICB)–0 (centre)
Shearer et al. (1988)	abs. travel times PKIKP; PKP(BC)	uniform	1.0
Shearer and Toy (1991)	diff. travel times PKP(BC)−PKIKP	uniform	0.9; data confined to IC top
Creager (1992)	diff. travel times PKP(BC)−PKIKP	uniform	3.5; data confined to IC top
Song and Helmberger (1993)	diff. travel times PKP(BC)−PKIKP	uniform	3.0
Tromp (1993)	normal mode splitting	radially varying	3.4 (ICB)–0 (centre)
Vinnik et al. (1994)	diff. travel times PKP(AB)−PKIKP	uniform	3.5; data isotropic IC top confined to IC centre
Shearer (1994)	abs. travel times PKIKP	layered model	3.5
Tromp (1995)	normal modes + travel times	radially varying	0.5–3.0
Su and Dziewonski (1995)	abs. travel times PKIKP	layered model	radius range [km] 920–620: 0.7 620–320: 1.6 320–0: 3.3
McSweeney et al. (1997)	diff. travel times PKP(BC)−PKIKP PKP(AB)−PKIKP	layered model	radius [km] 3.0 at 200 km 4.0 at 500 km weak below 500 km
Creager (1999)	diff. travel times PKP(BC)–PKIKP PKP(AB)–PKIKP	layered E–W model	radius range [km] 1220–600 0.5E 3.0W 600–0 3.0E 3.0W
Garcia and Souriau (2000)	abs. travel times PKIKP	layered E–W model	radius range [km] 1220–1120: 0.4E 0.7W 1120–820: 0.6E 3.3W 820–0: 2.6E 3.1W
Ishii and Dziewoński (2003)	abs. travel times PKIKP	layered model	radius range [km] 1220–300: 1.8 300–0: 3.7 tilted

Calvet et al. (2006)	abs. travel times PKIKP	layered model 2	radius range [km] 1220–440: 1.2 ± 0.1 440–0: 0.2 ± 0.3
		layered model 3	radius range [km] 1220–550: 1.6 ± 0.0 550–0: 2.0 ± 0.2
Leykam et al. (2010)	diff. travel times PKP(BC)–PKIKP excluding SSI data	uniform	0.7 ± 0.1; data confined to IC top
Irving and Deuss (2011)	diff. travel times PKP(BC)–PKIKP PKP(AB)–PKIKP	layered E-W model	0.5 ± 0.2E 4.8 ± 0.2W 1.4 ± 0.3E 4.5 ± 0.4W

Note: $^{a}\widehat{V_p}$ is cylindrical anisotropy strength expressed as a percentage. For an anisotropic IC with cylindrical symmetry and a fast axis aligned with the rotation axis of the Earth, this is equivalent to a P-wave speed perturbation (from the Earth reference model value) in the polar direction (parameter ϵ in Equation 4.3). This will be the maximum velocity perturbation (in the absolute sense) unless parameter σ (see Equation 4.3 and related text) is extremely large. For the models in which anisotropy strength varies with radius, the strength of anisotropy is expressed as a range.

multiplets (such as $_{10}S_2$ and $_{11}S_2$), are actually anomalously split singlets of the same multiplet ($_{10}S_2$) (Figure 4.5). It is these modes that led Masters and Gilbert (1981) to report that the IC must have a high Q. Based on the account and measurements of such, it was suggested that the weakly excited core modes could in fact be observed.

Aspherical Structure in the Core

Ritzwoller et al. (1986) reported that in fact approximately one-third of the observed multiplets splits are anomalously wide through the use of two new, independent techniques called 'singlet stripping' and 'non-linear regression'. Both techniques were developed to counterpoise the fact that the existence of aspherical structure within the Earth affects the low frequency response of normal modes in a non-linear way. The authors demonstrated that both techniques yielded robust fits to synthetic amplitude spectra and the distribution of singlets within a multiplet when assuming non-axisymmetric structure. These results supported the existence of deep, large-scale aspherical structure within the Earth's OC, because a departure from a spherically symmetric model at shallower depths of the Earth could not predict the measured, anomalously split multiplets. This was the first time that an asymmetric structure was inferred to exist in the deeper Earth shells, though its existence in the transition zone had already been proposed (Masters et al., 1982).

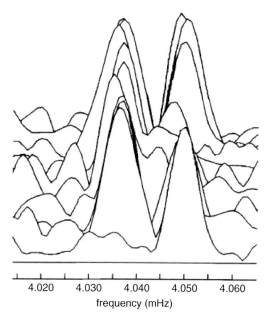

frequency (mHz)

Figure 4.5 Amplitude spectra of 13 recordings in the 50μ Hz band around $_{10}S_2$ following deep earthquakes from Tonga (22 June 1977) and the Banda Sea (22 June 1982). The equatorial stations are in back and the polar stations are in front. The only spheroidal mode in this band is $_{10}S_2$, and it is anomalously wide, spanning more than 14μ Hz. This is 8μ Hz more than predicted for a rotating, hydrostatic Earth. After Ritzwoller et al. (1986).

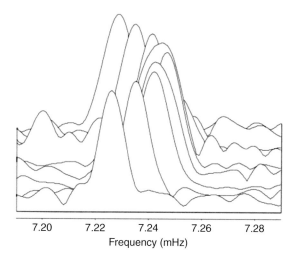

Frequency (mHz)

Figure 4.6 Amplitude spectra of singlet strips for the multiplet $_{18}S_4$. The strips are arranged in ascending azimuthal order. The singlet strip corresponding to the azimuthal order 4 is plotted in front. After Widmer et al. (1992).

At the time the work of Ritzwoller et al. (1986) was published, it was clear that in order to explain the measured anomalous normal mode splitting, either a substantial volume of heterogeneity of a non-axisymmetric character had to be present in the mantle (particularly, with a strong spherical C_2^0 component, indicating the dominance of zonal structure) or otherwise in the Earth's core along with large topography at the core boundaries. However, early tomography studies of the lower mantle (Dziewoński, 1984) did not support a strong C_2^0 anomaly near the CMB, which agreed with the measured splitting of the mantle-sensitive modes. Therefore, the focus was shifted to the core, and eventually the IC. This development was clearly outside the framework of spherically symmetric models of the Earth, and at the end of 1986 led two accompanying studies to propose a non-isotropic IC. We have seen in the previous section how Morelli et al. (1986) argued for the existence of IC anisotropy based on PKIKP travel time data. Here, we will see how the work of Woodhouse et al. (1986) supported an anisotropic IC.

There was growing evidence that anomalous PKIKP travel times and the splitting of normal modes sensitive to the deep Earth could be explained by an inhomogeneous IC. Nonetheless, Woodhouse et al. (1986) decided to consider a new possibility – that there is a simple anisotropic structure in the IC, symmetric with respect to the rotation axis of the Earth. The word 'simple' here refers to the most basic case of anisotropy, which would be a result of either a systematic orientation of crystals or a particular regime of convection in the IC. On an intuitive level, the dynamics of the core is influenced, if not dominated, by the Earth's rotation; therefore, it was reasonable to assume that anisotropy with a cylindrical symmetry about the Earth's rotation axis would fit the normal mode data. In simple terms, Woodhouse et al. (1986) compared the lowest harmonic degree expansions of Earth structure seen by normal modes with the structure theorised assuming anisotropic IC structure. They limited their analysis to the C_2^0 and C_4^0 terms of Earth structure. At first, the best fit solution was not satisfactory, as it predicted excessively large variations in travel times as compared to the findings of Morelli et al. (1986). This was somewhat remedied by reducing the modelled anisotropy with depth; this preserved the fit to the C_2^0 terms but sacrificed the fit to the C_4^0 terms (Figure 4.3c).

Thus, the original consideration of a directionally dependant IC velocity structure led to the establishment of one of the most prominent conceptual frameworks for IC studies in the last several decades. This pioneering seismological work was carried forward by other researchers, including mineral physicists and geodynamo modellers, who proposed a vast number of mechanisms and explanations for anisotropy in the IC (Figure 4.3).

4.3 Towards Complexity

4.3.1 First Contradictions and Resolutions

Naturally, the first research efforts that followed the pioneering work from the late 1980s focused mainly on how to confirm anisotropy. The most common goal was to determine the strength of anisotropy that could simultaneously fit body wave travel times and the splitting of normal modes. The researchers working at that time established a simple model of IC anisotropy that was a good first approximation for explaining body wave travel times. From there, additional travel time data targeting specific source-receiver paths were added to improve spatial coverage of the IC and further refine the proposed magnitude of IC anisotropy. This method allowed for a spatial variability in the IC anisotropy. Upon perturbing the model of IC anisotropy to fit the travel time measurements, its ability to fit the normal mode data is examined. If the fit is reasonably good, the model retains credibility. But what if it is not? Then there is a puzzling discrepancy, such as was encountered at the end of the 1980s and the beginning of the 1990s.

Discrepancy in Estimates of Anisotropy Strength from Travel Time Studies

Shearer et al. (1988) examined both PKP(BC) and PKIKP absolute travel times from the ISC catalogue and found no anomalous PKP(BC) travel times. This meant that any anomalous PKIKP−PKP(BC) travel time residuals must come from where the PKIKP and PKP(BC) ray paths differ, that is, the IC. However, they noted that their results were inconsistent with Morelli et al. (1986) in that shallow turning ray paths in the IC did not suggest anisotropy. They concluded that anisotropy is spread more evenly and has a strength of 1 per cent.

In a subsequent study, Shearer and Toy (1991) expanded their analysis to include PKP differential travel times. Interestingly, this was the first time these differential travel times were used to infer anisotropy in the IC since Cormier and Choy (1986) pioneered the technique to study IC heterogeneities several years earlier. An example of a two-step method of measuring PKP(AB)−PKIKP differential travel times using the Hilbert transform operator and cross-correlation is shown in Section 2.4 in Figures 2.5 and 2.6. Shearer and Toy (1991) estimated a uniform anisotropy strength of 0.6–1.2%, thus confirming their previous prediction that IC anisotropy is about three times weaker than proposed in the pioneering work of Morelli et al. (1986). It is worth noting that at the time their dataset consisted of a single polar path. They also confirmed the existence of aspherical isotropic velocity structure, but because of insufficient data coverage argued for a trade-off between lateral heterogeneity and anisotropy.

PKP(BC)−PKIKP and PKP(AB)−PKIKP differential travel time residuals can be plotted as a function of angle ξ. We can then study data trends and compare them with the predictions after accounting for various phenomena that might effect travel times, such as heterogeneity or anisotropy. Two such global datasets from the study of Tkalčić et al. (2002) are shown in Figure 4.7, with the geographical location of selected sources and receivers forming polar ($\xi \leq 35°$) paths indicated. It is immediately visible that the PKP(BC)−PKIKP dataset (Figure 4.7a) has a smaller scatter than the PKP(AB)−PKIKP dataset (Figure 4.7b). Polar angles display large scatter, with some clearly anomalous differential travel time residuals in both datasets. Most anomalous ray paths correspond to the South Sandwich Islands earthquakes recorded in Alaska. Figure captions contain more detailed information about each marked source-receiver path.

The work of Creager (1992) was also among the first studies of IC anisotropy using differential travel times and utilised waveforms available from the GDSN (Global Digital Seismographic Network). Previously, Shearer and Toy (1991) used this resource as well, but only for a single polar path. In contrast the new study of 1992 featured about 20 polar paths. This study estimated IC anisotropy to be approximately 3.5 per cent in the lower latitudes of the upper IC. Significantly, Creager (1992) also found that large-scale heterogeneity in isotropic velocity had variations on the order of 0.5 s. Song and Helmberger (1993) studied more polar paths using earthquake and explosion recordings from the WWSSN (Worldwide Standardised Seismograph Network) and confirmed the finding of Creager (1992) about the anisotropy strength. The differential travel time residual predictions from this model are compared with the collection of global PKP(AB)−PKIKP and PKP(BC)−PKIKP differential travel time data measured through cross-correlation (Tkalčić et al., 2002) (Figure 4.8c,d). While the general trend observed in the measured data is fit reasonably well, the scatter in data prevents a higher variance reduction. We will return to this figure once more in Section 4.3.4.

At the beginning of the 1990s there were glaring discrepancies between estimates of IC anisotropy strength. This was not only the case among travel time studies, but also among normal mode studies, as we will see in the next subsection. Motivated by this problem, Shearer (1994) reconsidered the analysis of PKIKP data from the ISC catalogue, now empowered with novel averaging techniques. The method amended the previous oversight of selecting too narrow of windows used for binning and averaging the large amounts of travel time data from the ISC catalogue. He found IC anisotropy to be as strong as claimed by Creager (1992) and Song and Helmberger (1993) and concluded that the top 50 km of the IC is isotropic.

Discrepancy in Estimates of Anisotropy Strength from Normal Mode Studies

To clarify the concept of finding a model of IC anisotropy that simultaneously fits both travel time and normal mode data, imagine that we must carve an abstract wooden object for which we have a previously made box. At the beginning, the

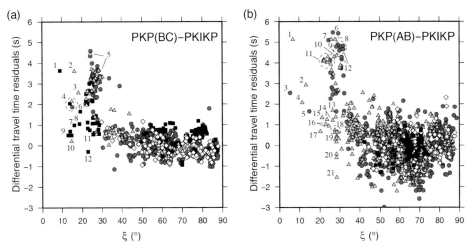

Figure 4.7 Differential travel time residuals plotted as a function of the angle ξ between the PKIKP leg in the IC and Earth's spin axis. All differential travel time residuals are calculated (taking standard ellipticity corrections into account) with respect to model ak135 (Kennett et al., 1995). All earthquake locations and origin times are corrected with respect to the relocation catalogue of Engdahl et al. (1998). (a) PKP(AB)−PKIKP residuals. Different symbols correspond to data from four different analyst sources: circles, McSweeney et al. (1997); squares, Tanaka and Hamaguchi (1997); diamonds, *Souriau*, personal communication; triangles, Tkalčić (2001). Some specific polar paths are indicated by numbers as follows: (1) Novaya Zemlya to SNA (Antarctica); (2) Alaska to SPA (Antarctica); (3) South Sandwich Islands (SSI) to MBC (NW Territories, Canada); (4) Alaska and north Canada to SPA; (5) SSI to northeast Asia and Alaska; (6) (52S, 140E) to NOR (Greenland, Denmark); (7) Siberia to SBA (Antarctica); (8) (53S,160E) to NOR; (62N,154E), (64N, 125E) and (60N, 169E) to SPA; (10) (60N, 153W) to SPA; (11) various locations to SYO (Antarctica); (12) south of Australia to NOR. (b) PKP(BC)−PKIKP residuals. Different symbols correspond to data from four different analysts sources: circles, McSweeney et al. (1997); triangles, Tkalčić (2001); diamonds, *Souriau*, personal communication; squares, *Wysession*, personal communication. Some specific polar paths are indicated by numbers as follows: (1) Svalbard Sea to SPA; (2) (63N, 143W) to SPA; (3) (79N, 124E) to SPA; (4) (67N, 173W) to SPA; (5) (60N, 153W) to SPA; (6) mid-Atlantic ridge to Alaska; (7) SSI to COL (Alaska, USA); (8) SSI to BILL (Siberia, Russia); (9) SSI to SEY (Siberia, Russia); (10) southeast Pacific to NRIL (Siberia, Russia); (11) SSI to COL and NRIL; (12) mid-Atlantic ridge to Alaska; (13) south of New Zealand to FRB (NW Territories, Canada); (14) Bouvet Islands to COLA (Alaska, USA) and INK (NW Territories, Canada); (15) Aleutian Islands to SYO; (16) Sea of Okhotsk to SYO; (17) south of New Zealand to KBS (Svalbard, Norway); (18) various locations to SYO; (19–21) Aleutian Islands to SYO. Modified from Tkalčić et al. (2002).

piece of wood we have for carving is bigger than our box. The requirements are that the wooden piece be carved into an object that fits into the box, but that it is not so small as to be loose inside. In other words, the object's shape does not matter much, as long as it can fit tightly into the box. All we need to do is carve it until, upon a convenient rotation, it can be enclosed in the box.

Now, imagine that we reverse the above process and instead want to figure out if a given box has an object inside it or not. We know the mass of the box when it is empty. The most 'general check' would be to weigh the box to deduce if an object is inside. If the mass of the box is bigger than that of an empty box, we know that an object is inside, but we would not know anything about its shape. On an intuitive level, we can think of the fitting of normal mode spectra as a 'general check' as in the above example, and of the fitting of travel times as a 'fine tuning' tool that we would somehow use to predict the object's shape. There is a parallel here with the Earth if we think of the box from the above example to represent the Earth and if we think of the object's shape as representing Earth's structure. Indeed, a normal mode 'senses' Earth structure as an integral over the entire Earth's radius. If its sensitivity is uniformly spread over the Earth's radius, we can immediately see that this mode alone will not yield sufficient information about how the structure is distributed along the radius. We could conclude something general, e.g. its mean density, but we could not say anything about its spatial variation.

Fortunately, the sensitivity of each normal mode is not uniform along the radius. We can theoretically predict the shape of sensitivity of each mode as a function of Earth's radius. For example, some normal modes are not sensitive to IC structure at all. They simply do not sense so deep. Others have a small percentage of their elastic energy sensitive to the top of the IC. Only a few modes sense all the way to the Earth's centre (see Figure 2.15). Hence, if IC anisotropy exists and is uniform across the IC volume, it contributes significantly to the splitting of all normal modes sensitive to the IC, regardless of the depth at which their sensitivity in the IC peaks. Those modes with more energy in the IC will naturally be more effected. For modes sensitive only to the outer part of the IC, significantly split spectra would indicate strong anisotropy in the shallow IC.

Li et al. (1991) and Widmer et al. (1992) challenged the IC anisotropy hypothesis, mainly because with this phenomenon alone they were not able to explain certain normal mode splitting measurements. Instead, they argued that anomalous structure in the OC of harmonic degree $s = 2$ and with a density facilitating hydrostatic equilibrium can explain a large amount of the normal mode data, albeit not without the cost of requiring an explanation for the close relationship between P wave velocity and density. More about this is written in Section 4.3.4. Subsequently, Tromp (1993) assumed a radial dependence of IC anisotropy. He showed that the fit to a set of splitting measurements improves significantly when the effects of IC anisotropy are included in comparison to the case when only rotation and ellipticity effects are considered. He recognised that lateral heterogeneity contributes to splitting, so his estimate of 3 per cent should be considered an upper limit to IC anisotropy strength.

Radially Variant Versus Axisymmetric Anisotropy in the IC

After the Bolivia and Kuril Islands events, however, Tromp (1995) explained a significant portion of both the normal mode splitting and PKP differential travel time data with a radially dependent cylindrical model of anisotropy. This model was constrained in its deepest part exclusively by travel time observations, as there are few normal modes with sensitivity near Earth's centre. Consequently, increasing the magnitude of IC anisotropy near the centre has little effect on normal modes. A comparison between the differential travel time predictions based on this model with the collection of global PKP(AB)−PKIKP and PKP(BC)−PKIKP differential travel time data measured through cross-correlation (Tkalčić et al., 2002) is shown in Figure 4.8a, b. As with other models of cylindrical anisotropy in the IC, the scatter in data prevents further variance reduction (we will return to this comparison once more in Section 4.3.4). Although the model explained most of normal mode splitting data reasonably well, part of the signal (e.g. the anomalous splitting of $_{13}S_2$) was not yet adequately explained.

Despite the contradicting strength estimates of IC = anisotropy in the first decade since its discovery, most authors by the mid 1990s argued that the strength of anisotropy was well ascertained. The existence of liquid inclusions was considered unlikely based on independent observations of anisotropy in attenuation (Souriau and Romanowicz, 1996) and the overall agreement between mineral physics and seismological observations of 3–3.5% anisotropy. Thus the central question was whether the cause of alignment amongst IC iron crystals is convection (Jeanloz and Wenk, 1988) or freezing by the magnetic field (Karato, 2003). An accurate answer necessitated the modelling of IC anisotropy in 3D.

Romanowicz et al. (1996) and Durek and Romanowicz (1999) used the formalism introduced by Widmer et al. (1991) to represent a departure from radial symmetry that created patterns associated with simple physical processes in the IC. The models they obtained were characterised by two zones of significant anisotropy: a central zone parallel to the Earth's rotation axis and a superficial zone in the equatorial region (Figure 4.3f). They recognised that these models fit the inversion data better than models with a radial symmetry. The resulting picture was compatible with large-scale convection but incompatible with 'frozen in' anisotropy, which should preserve radial symmetry. The fit to normal mode amplitude and phase spectra without and with IC anisotropy is shown in Figure 4.9.

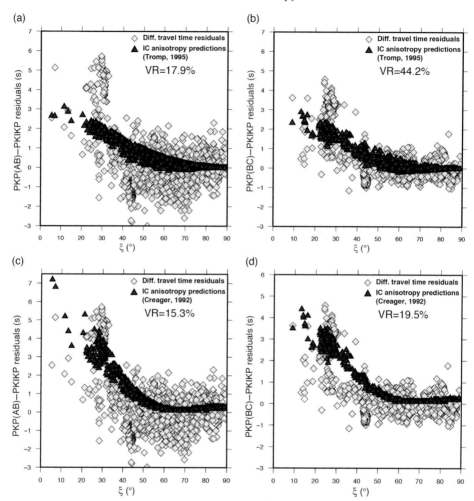

Figure 4.8 Measured (diamonds) versus theoretically predicted (triangles) differential travel times plotted as a function of the angle between the PKIKP ray path in the IC and the Earth's rotation axis (ξ). Comparison of measured differential travel times and theoretical predictions based on the IC anisotropy model of Tromp (1995) for the (a) PKP(AB)−PKIKP dataset and (b) PKP(BC)−PKIKP dataset. Comparison of measured differential travel times and theoretical predictions based on the IC anisotropy model of Creager (1992) for the (c) PKP(AB)−PKIKP dataset and (d) PKP(BC)−PKIKP dataset.

After more normal mode splitting measurements became available from the great Bolivia and Kuril Islands earthquakes in 1994, it became clear that IC anisotropy is not simple. These new measurements required IC anisotropy to vary with depth or even in three dimensions. We have seen in Section 4.2.1 that Morelli et al. (1986) and Woodhouse et al. (1986) speculated about the depth dependence

mode 3 S 2

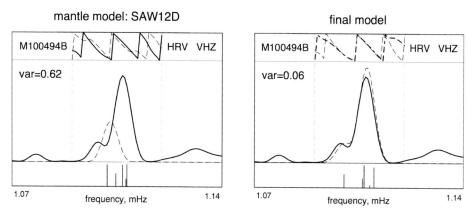

mode 13 S 2

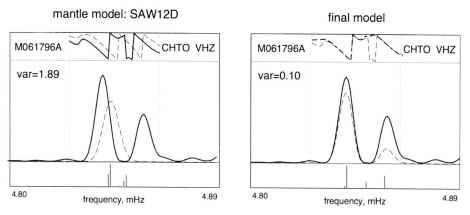

Figure 4.9 Comparison of the observed (solid lines) and predicted (dashed lines) spectra for IC sensitive normal modes $_3S_2$ (top row) and $_{13}S_2$ (bottom row). Each frame contains the phase (top panel) and amplitude (bottom panel) spectra. Vertical bars in the bottom panels indicate the frequencies and relative amplitudes of the singlets contributing to the theoretical spectra. The predictions of the retrieved splitting functions (right) are significantly improved compared to those based only on mantle structure, ellipticity and rotation (left). After Durek and Romanowicz (1999).

of IC anisotropy since their simple models of cylindrical IC anisotropy were not unique nor did they fit all data equally well.

Naturally, the problem of simultaneously fitting the travel time and normal mode data continued to intrigue seismologists and provoked a number of studies that used new ray paths sampling at different directions and different depths through the IC. This process can be thought of as the 'fine tuning' mentioned in our thought experiment involving a wooden object and its box. Travel times can determine Earth structure in the same way that sonar could determine the shape of our enclosed object.

Song and Helmberger (1993) studied the ray paths from Novaya Zemlya to the station Scott Base (SBA) in Antarctica. They concluded that IC anisotropy must be non-uniform with depth, and more specifically, stronger near the top of the IC and weaker towards the IC centre, because for that particular path the travel time residuals were smaller than predicted based on a uniform 3.5 per cent IC anisotropy model. The study of Vinnik et al. (1994) featured the longest ray path through the Earth using the South Pacific earthquakes recorded at the GEOSCOPE station SEY. At those large epicentral distances ($\Delta \geq 165°$), the PKP(BC) phase does not exist, and the differential travel time measurement is made between the PKIKP and PKP(AB) phase arrivals. The deeper IC sampling comes at the cost of less similar ray paths in the mantle. They determined that an IC anisotropy of 3.5 per cent fit the newly observed travel time data (Figure 4.3d). But simply extrapolating this strong value of anisotropy throughout the entire IC directly contradicts the research of Tromp (1993), which revealed that certain normal mode measurements contradicted strong anisotropy towards the IC centre.

4.3.2 Isotropic Outermost IC?

Studies That Led to the Concept of Isotropic Top of the IC

Seismological   studies of PKIKP travel times in the late 1980s and early 1990s inferred that the top of the IC is isotropic (Shearer et al., 1988; Shearer and Toy, 1991; Shearer, 1994), though this was not immediately recognised in other studies of IC anisotropy (e.g. Creager, 1992; Tromp, 1993). It was clear that the IC was not uniform: a homogeneous, isotropic layer encased an anisotropic heart. Though the nature of the transition was unknown and the exact thickness of the isotropic layer left uninvestigated in early studies. The origin of such a layer could be related to dynamic processes near the ICB which would prevent an anisotropic fabric to develop or subsist. Subsequent studies, however, not only confirm the existence of an outer isotropic layer, but additionally determine that its thickness varies longitudinally.

Using seismic waveform modelling tools, Song and Helmberger (1995) studied the waveforms sensitive to the uppermost part of the IC, in the triplication range, which are normally excluded from the ISC catalogue due to the difficulties associated with obtaining reliable picks. The authors remedied the problem by directly comparing synthetic seismograms with recorded seismograms, and concluded that the top 150 km of the IC is only weakly anisotropic (≤ 1 per cent), whereas the top

60 km is likely completely isotropic, thus confirming earlier results (e.g. Shearer, 1994).

Undeterred by claims of uniform IC anisotropy (McSweeney et al., 1997), the authors eventually hypothesised that the IC has a laterally varying transition zone about 100−250 km below the ICB where the compressional wave velocity jumps in amplitude (Song and Helmberger, 1998) (Figure 4.3h). For a seismological study to claim the existence of such a velocity jump within the Earth, it is crucial to explain its effect on seismic waves and confirm its observation. A transition zone less than 50 km thick would likely create a reflection strong enough to be recorded on short period seismograms and a broadening of PKIKP pulses that would be recorded on broadband seismograms. The problem was that this effect was not observed with sufficient consistency as to confirm the ubiquitous presence of a transition zone of unvarying depth. Based on global observations of polar ray paths through the IC and waveform modelling, Song and Helmberger (1998) proposed a new view of IC structure where the transition zone, separating an upper, isotropic layer from a lower, anisotropic layer, varied laterally.

Not surprisingly, there is a strong trade off between the transition zone depth and the velocity jump (i.e. moving the depth of the transition zone from 200 to 300 km would require a change in velocity jump from 3.5 to 5 per cent). Another problem associated with a laterally varying transition zone is that if it represents a progression from one crystallographic structure of iron to another, the depth variation could not be easily justified as IC pressure and temperature are assumed laterally consistent. Therefore, a physical mechanism to explain this model remained elusive.

Garcia and Souriau (2000) further constrained the isotropic layer by performing a stochastic analysis of a combined ISC and hand-picked differential travel time (PKP(BC)−PKIKP) dataset that allowed for lateral variation in the isotropic layer thickness. Their IC model exhibits uniform anisotropy of 2 to 3 per cent encased in a uniform isotropic layer of varying thickness: 400 km in the eastern hemisphere and 100 to 200 km in the western hemisphere (Figure 4.3i). In a ensuing study, Garcia (2002) confirmed the model of Garcia and Souriau (2000). The waveform inversion supported an isotropic outer IC with a hemispherical velocity variation that correlated with the hemispherical variation in anisotropic velocity 100 km beneath the ICB. Garcia (2002) emphasises a connection between higher P-wave velocities and the absence of anisotropy in the eastern hemisphere and between lower P-wave velocities and the presence of anisotropy in the western hemisphere. He argues that such correlation can be explained by anisotropy formation by grain boundary migration in the presence of heterogeneous heat flow at the ICB, as proposed by Yoshida et al. (1996).

Difficulties Associated with Radial Distribution of Anisotropy Strength

Approximately concurrent studies of differential PKP(AB)−PKIKP travel times taken from the ISC catalogue (Su and Dziewonski, 1995) and measured by cross-correlation from WWSSN analogue seismograms (Song, 1996) confirmed strong central IC anisotropy of 3 to 3.5 per cent, thus confirming earlier results of Vinnik et al. (1994). A common justification was the significantly higher variation between polar and equatorial paths in the absolute PKIKP travel times than in the corresponding PKP(AB) travel times. This was interpreted as proof that the phenomenon effecting the travel times must be in the IC. We will return to the study of Su and Dziewonski (1995) in Section 4.3.3 when discussing the lateral variation of IC anisotropy.

McSweeney et al. (1997) tripled the number of PKP differential travel times using the cross-correlation technique and found the magnitude of IC anisotropy to be 3 per cent about 200 km below the ICB, with an increase to 4 per cent about 500 km below the ICB. Below this depth resolution is reduced, but here they speculated that anisotropy must weaken. Again, the best fitting direction for the fast axis of anisotropy was close to the rotation axis of the Earth. This study recognised that the most serious problem with reconciling 3 to 4 per cent anisotropy with theoretical calculations from first principles (Stixrude and Cohen, 1995) lies in the fact that iron crystals would have to be 100 per cent aligned.

The realisation that the outer part of the IC is isotropic was somewhat a surprising turn in the research of IC anisotropy, attesting to the complicated relationship between our ability to measure the magnitude of IC anisotropy and its radial dependence. It complicated the fits to normal modes even further, because some modes have their entire sensitivity in the IC exclusively in its 'ceiling', under the ICB. For example, such modes are $_{11}S_5$ and $_{16}S_6$. If there is no anisotropy in the uppermost shell of the IC, how can the anomalous splitting of these modes be explained? In fact, most modes require the top part of the IC to be quite anisotropic, because their sensitivity steeply drops down in the deeper IC.

To summarise, there were several serious difficulties associated with the concept of IC anisotropy in the late 1990s: a) seismological studies of the IC using travel times contradicted each other, b) an apparent discrepancy existed between normal mode splitting and differential travel time data, c) the required magnitude of anisotropy was too large, and d) the physical mechanism behind radial and lateral variation of IC anisotropy was unknown. To expound a little further, the first problem was that while some studies argued for strong anisotropy persisting down to the Earth's centre (e.g. Vinnik et al., 1994; Song and Helmberger, 1998), others proposed a weakening in the centremost part of the Earth (McSweeney et al., 1997). The second problem was that fitting travel time measurements required an isotropic outer IC, which contradicted IC models that fit normal mode splitting measurements. The third problem was that the proposed magnitude of IC anisotropy required that the IC iron crystals be perfectly, and therefore unrealistically, aligned. The last but not least problem was that a physical mechanism that would explain lateral variation of the postulated transition zone between the isotropic and anisotropic part of the IC had not been found. As we will see in the last section of this chapter, this issue led to a series of papers challenging the existence of IC anisotropy.

4.3.3 Longitudinal Variations in IC Anisotropy

By the end of 1990s, it became clear that various seismological data are difficult to reconcile using a simple, cylindrical IC anisotropy model. Su and Dziewonski (1995) supported this conclusion by highlighting the limitations of such a model when explaining International Seismological Centre (ISC) travel time data. Before them, Shearer (1994) and Creager (1992) addressed longitudinal variation by plotting absolute PKIKP travel times as a function of longitude and ray azimuth on cylindrical projections, or as a function of longitude and angle ξ on polar projections, though the focus of the work was on the depth variation of anisotropy.

Thus, Su and Dziewonski (1995) were the first to more thoroughly explore longitudinal anisotropy variation and its potential explanations. They also observed variation with depth by parameterising their model with four layers of equal thicknesses. They introduced another parameter – that of varying cylindrical anisotropy in space (Figure 4.3e). Their best fit model featured a tilt of the fast axis of anisotropy with respect to the rotation axis of the Earth. For the longitudinal variations at different depths, they used a Taylor series expansion to observe differences in magnitude. However, their chief goal was to find the best uniform cylindrical anisotropy model.

Tanaka and Hamaguchi (1997) took the study of longitudinal variations a step further. We discussed in Chapter 3 how they established a hemispherical pattern to the IC isotropic velocity structure. They took a unique two-step approach: firstly, they isolated the 'equatorial dataset' (a travel time dataset comprising PKIKP paths which sample the IC in the quasi-equatorial plane) to estimate isotropic effects; secondly, they considered quasi-polar paths to observe anisotropic effects. Namely, if anisotropy is cylindrical with a strong effect on quasi-polar paths and only negligible effect on quasi-equatorial paths, separating the former from the latter subset will efficiently remove the trade-off between isotropic and anisotropic velocity.

After isotropic velocity in the IC was shown to be hemispherically variant, it remained to study polar paths and determine if this was also the case for anisotropy.

Unfortunately, at the time of the study by Tanaka and Hamaguchi (1997), they only had quasi-polar paths for a limited number of events, particularly in the qWH where only SSI events were available. They hypothesised that the hemispherical pattern in the IC included both the isotropic to anisotropic velocities. Since they did not have data to contradict this hypothesis, they accepted it (Figure 4.3g). The PKP(BC)−PKIKP differential travel time residuals from their study, now divided between the two hemispheres (East: 43°E–177°E; West: 183°W–43°E) could be compared with anisotropy predictions from existing models (Figure 4.10. Using the same hemisphere definitions, Tkalčić et al. (2002) investigated the possible effects of lowermost mantle structure and IC anisotropy on expanded, global datasets of PKP(AB)−PKIKP and PKP(BC)−PKIKP differential travel time residuals and confirmed the hemispherical dichotomy in isotropic velocities for the PKP(BC)−PKIKP dataset, but not for the PKP(AB)−PKIKP dataset (Figure 4.11). At polar sampling angles, only the western hemisphere data exhibit large anomalies.

The study of Tanaka and Hamaguchi (1997) remained relatively unnoticed by geodynamicists until the work of Niu and Wen (2001) provided confirmation using differential PKiKP−PKIKP travel times. Thus, the newly observed hemispherical pattern of IC anisotropy and the lack of a physical mechanism to explain it by added to the problems listed at the end of Section 4.3.2 and provoked seismologists

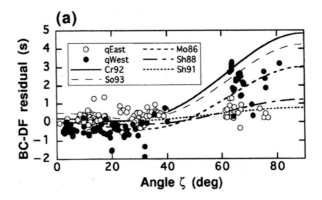

Figure 4.10 PKP(BC)−PKIKP residuals (note that PKIKP waves are sometimes referred to as PKP(DF) waves) as a function of angle ζ between the PKIKP ray and the equatorial plane ($\xi = 90° - \zeta$). Open and solid circles represent residuals located in the quasi-eastern (43°E–177°E) and quasi-western (183°W–43°E) hemispheres, respectively. Predicted PKP(BC)−PKIKP differential travel time residuals from various IC anisotropy models are shown by different lines according to the figure legend. Abbreviations are as follows: Mo86 (Morelli et al., 1986); Sh88 (Shearer et al., 1988); Sh91 (Shearer and Toy, 1991); Cr92 (Creager, 1992); So93 (Song and Helmberger, 1993). After Tanaka and Hamaguchi (1997).

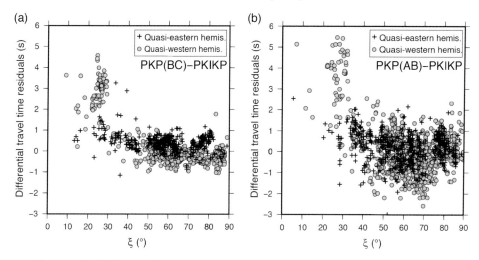

Figure 4.11 Differential travel time residuals plotted as a function of the angle ξ between the PKIKP leg in the IC and Earth's spin axis. All differential travel time residuals are calculated (taking standard ellipticity corrections into account) with respect to model ak135 (Kennett et al., 1995). All earthquake locations and origin times are corrected with respect to the relocation catalogue of Engdahl et al. (1998). Data are divided into two subsets according to the PKIKP ray bottoming point longitudes in the IC following the definition from Tanaka and Hamaguchi (1997). Pluses are qEH data points (43°E–177°E). Circles are qWH data points (183°W–43°E). (a) PKP(BC)−PKIKP residuals; (b) PKP(AB)−PKIKP residuals. Modified from Tkalčić et al. (2002).

to seek alternative explanations to IC anisotropy, outside the IC. These studies will be the topic of the next section.

4.3.4 Alternative Hypotheses

Imagine that we have a thousand travel time data points measured by the global seismological network (this is close enough to the number of available measurements in the late 1990s). We first plot the travel time residuals as a function of the angle between the PKIKP wave ray paths and the rotation axis of Earth (Figure 4.4). We then fit a curve to these data points as to minimise the misfit. The type of function that satisfies this criterion allows us to hypothesise about the physical phenomenon causing the travel time variations. Recall that a quadratic function of angle ξ minimised the misfit in travel time data and agrees with a model of transverse isotropy (Equation 4.2; Appendix C). A promising hypothesis is one which does not contradict the data. Recall that it was this lack of contradiction that led Lehmann (1936) to propose the existence of the IC. However, as the number of measurements grew, so did the evidence against a simple cylindrical model of anisotropy. In the next section we will see that in fact seismologists who studied the PKP travel times started to question the existence of IC anisotropy entirely.

To Bin or Not to Bin?

In a physics experiment it is often advantageous to average multiple measurements of the same physical property. For example, Cavendish (1798) designed an experiment that led to the estimate of the gravitational constant. Upon repeating such an experiment, each time with the same masses of balls and length of strings, we could compute the mean of the force moments and their standard deviations before estimating the gravitational constant. To get more sophisticated, we could change the apparatus setting, e.g. by increasing the mass of the balls, and plot the changes in one parameter as a function of another. For example, we could plot the change in the force moment as a function of the mass of the balls. The data could be binned according to different masses, with each bin characterised by the mean and standard deviation. This should lead to a reasonable estimate of the gravitational constant provided our experimental apparatus is free from outside gravitational influences. As the name suggests, the gravitational constant is a constant quantity, so in theory, regardless of the masses we use, the length of the springs, the orientation of our apparatus, or the geographic location of our lab, we should obtain approximately the same result.

To translate the above physics experiment to the case of IC anisotropy, let us compare the gravitational constant with the magnitude of IC cylindrical anisotropy, with the axis of symmetry aligned with the rotation axis of Earth. The different ball masses used in the experiment represent different angles of IC sampling. The angle of sampling effects travel times. If the ray path is parallel to the slow axis, the observed anisotropy effect on travel times will be zero. If, on the other hand, the ray path is parallel to the fast axis, the observed effect will be maximal. We could, eventually, deduce the magnitude of IC anisotropy from the observed changes in travel times as a function of angle ξ. Theoretically, the magnitude difference between the two extreme paths is the magnitude of anisotropy (this is featured in the last column of Table 4.2).

If we assume that IC anisotropy is cylindrical, with its fast axis perfectly aligned with the rotation axis of Earth, we have a simple working hypothesis with which we can approach the measurement anisotropy magnitude. Due to the assumed cylindrical symmetry, we could consider each measurement performed at the same angle ξ to be a repeated measurement, regardless of the geographical co-ordinates of the corresponding ray path. In other words, the travel time measurements from the South Sandwich Island earthquakes recorded in Alaska would be binned with the travel time measurements from any source location in the northern hemisphere and recorded in Antarctica as long as the angle ξ is the same (and the ray segment in the

Table 4.3 *PKIKP ray path characteristics in the IC for an increasing epicentral distance in model ak135 for a source at 0 km depth.*

Epicentral distance [°]	Bottoming depth below ICB [km]	Ray length in IC [km]	Time spent in IC [s]	Fraction time in IC [%]
145	125.6	1052.5	94.8	8.0
150	217.7	1365.5	122.9	10.4
155	346.9	1681.9	151.0	12.6
160	500.9	1954.6	175.1	14.6
165	671.7	2170.0	194.0	16.1
170	850.7	2321.8	207.2	17.1
175	1035.1	2412.1	215.1	17.8
180	1221.0	2442.0	217.7	18.0

IC is of the same length). Table 4.3 shows the characteristic parameters of PKIKP rays in the IC. For example, we can see how the PKIKP leg length and travel time spent in the IC relate to epicentral distance.

We could more loosely bin the data so that we group differential travel time measurements of the rays with similar geometries together. To simplify the problem, we could bin the data in nine bins, each spanning $10°$ (recall that the angle ξ is defined between $0°$ and $90°$). We could then compute the mean and standard deviation for each bin and invert for the best fitting curve through our nine points to find the magnitude of anisotropy. This could be the end of the story.

However, what if our main assumption that the IC anisotropy is cylindrical is wrong and anisotropy is more complex, with its magnitude varying longitudinally or its fast axis out of alignment with the rotation axis of the Earth? In that case, the logics behind data binning with respect to the angle ξ would be completely wrong, yielding flawed estimates for the magnitude of IC anisotropy. The data points within a given bin would not correspond to the measurements of a cylindrical anisotropy, but to something else, and their mean and standard deviation would not be interpretable. Moreover, the geographical location of the measurements would now not only be relevant, but crucial.

IC anisotropy questioned using travel time data

The plots of travel times residuals as a function of the angle ξ showed considerable scatter (Tkalčić, 2001; Figure 4.12). Within the same bin, the scatter was larger than what any reasonable model of cylindrical IC anisotropy could predict, even if the depth variation of anisotropy magnitude was taken into account. Some polar data

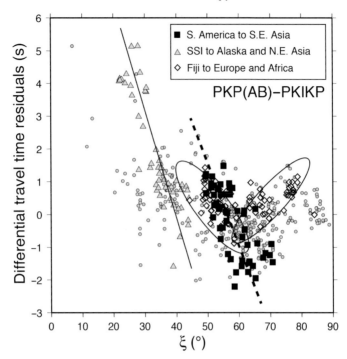

Figure 4.12 PKP(AB)−PKIKP differential travel time residuals plotted as a function of the angle ξ between the PKIKP ray path and Earth's spin axis. Data are binned in three groups based on the geographical locations of the sources and receivers (explained in the legend). Other data are shown by circles. Distinct trends are visible for all three datasets and are marked by a solid line (SSI to Alaska and northeast Asia), dashed line (South America to southeast Asia) and two ellipses (Fiji Islands to Europe and Africa). Modified from Tkalčić (2001).

were not largely anomalous, which contradicted a cylindrical anisotropy model with the fast axis parallel to the Earth's rotation axis. The subsets of data bottoming in the IC at different locations fit different quadratic curves, which suggested that transverse isotropy in the IC is not a globally consistent phenomenon. The scatter in travel time data could be explained by a more complex symmetry model, for example if the magnitude varied longitudinally. But even if this were the case, it was puzzling why only the SSI data recorded in Alaska produced such scattered data that the shape of the travel time residuals as a function of angle ξ looked like the 'letter L' rather than a quadratic cosine function.

The above considerations added to an already substantial set of issues regarding both travel time and normal mode observations. This led Bréger et al. (1999) and Bréger et al. (2000) to consider the possibility that the observed variations in travel times did not actually originate from IC anisotropy (Figure 4.3j). Despite the similarity of the ray path associated with each differential travel time, the observed scatter in the travel time data could instead be a consequence of lateral variations near the source, in the lowermost mantle, in the OC, or in the IC.

The initial assumption in many studies is that the crust and mantle contribution to travel times is avoided by the use of differential travel time analysis. This is not entirely true however, especially at the frequencies at which PKP body waves are usually analysed (about 1 Hz). Figure 4.13 illustrates the severity of the mantle heterogeneity effect on the PKP(AB)−PKIKP differential travel times. The rays are traced through the Earth using the global P-wave velocity model of Kárason and van der Hilst (2001) to correct PKIKP and PKP(AB) travel times as predicted by model ak135 (Kennett et al., 1995). For the SSI to Alaska geometry, the effect of the mantle is opposite on the PKP(AB) and PKIKP: nearly two seconds are added to PKP(AB) wave arrival times, while PKIKP waves speed up by a bit more than one second (Figure 4.13a). Subduction zones on both ends of this path and pronounced slow material in the mantle in the south Atlantic and spanning the entire mantle, significantly delay the PKP(AB) arrival and advance the PKIKP arrival (Figure 4.13b). Bréger et al. (2000) demonstrated that the scatter observed in PKP(AB)−PKIKP differential travel time data can be explained entirely by deep mantle structure. They hinted that the effects of mantle structure might perturb predicted differential travel time data to the point that all conclusions made about IC anisotropy, particularly in its central part, could be erroneous.

While the PKIKP and PKP(BC) ray paths are closer in spatial proximity, and therefore grant more reliable differential travel time measurements, than the PKP(AB) and PKIKP ray paths, the separation is not negligible. Even when travel times are corrected for mantle structure, the mantle is still a significant source of noise for core-sensitive phases since mantle structure is less understood in the lowermost mantle. Additionally, Helffrich and Sacks (1994) showed that PKP differential travel times can have significant azimuthal travel time anomalies consistent with a near-source slab effect. They argued that large biases may arise in PKP differential travel times depending on the slab geometry, velocity, and slowness difference between the two phases.

Bréger et al. (1999) analysed a global set of PKP(BC)−PKIKP travel time residuals (with respect to ak135) by performing a simple experiment. They traced the ray paths of PKP(BC) and PKIKP through the existing mantle models and corrected for heterogeneity, which explained a large amount of the signal in the differential travel times. The rapid variations seen in PKP(BC)−PKIKP (on scales shorter than several hundred kilometres) could thus be interpreted either as rapid variations in IC anisotropic properties or as rapid variations of velocity in the lowermost mantle. A combination of both seemed the likely scenario given the strong correlation between the features in the existing tomographic models (even though

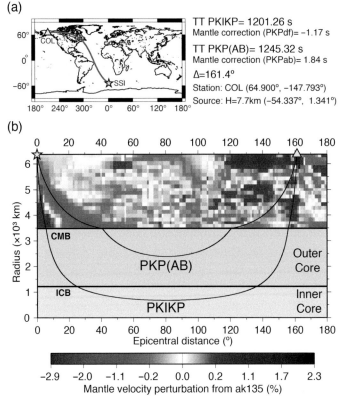

Figure 4.13 Illustration of the effect of mantle heterogeneity on PKP(AB)−PKIKP differential travel times. (a) Surface projection for a ray path from an earthquake in the South Sandwich Islands region to station College, Alaska (red line). Source and receiver information (star and triangle in the map), predicted total travel times through the Earth from ak135 model (Kennett et al., 1995) and corresponding corrections for mantle structure using the P-wave tomography model of Kárason and van der Hilst (2001) are given on the right. (b) Vertical cross-section through the Earth with major discontinuities (CMB: core-mantle boundary; ICB: inner core boundary) along the ray path shown in (a). Ray paths of PKP(AB) and PKIKP (black lines) through P-wave tomography model of Kárason and van der Hilst (2001). P-wave velocity perturbations from model ak135 are shown at the bottom. Modified from Tkalčić (2001). (A black and white version of this figure will appear in some formats. For the colour version, please refer to the plate section.)

the P-wave velocities were scaled up from the S-wave velocities) and the measured travel time residuals. Bréger et al. (1999) acknowledged that they could not distinguish between the slab effects, such as demonstrated by Helffrich and Sacks (1994), and the effect of lowermost mantle heterogeneity, but the main point was that IC anisotropy was not needed to explain anomalous PKP travel times. Bréger et al. (2000) analysed a global set of PKP(AB)−PKIKP travel time residuals (with respect to ak135) by tracing the ray paths of PKP(AB) and PKIKP through the existing mantle models and explained differential travel times by lowermost mantle structure only.

The assertion that differential travel times are affected by mantle structure was based on the observation of different scale heterogeneities in the lowermost mantle and coherent patterns in travel-time residuals in relation to the geographical locations of the associated earthquakes (Tkalčić, 2001). Tkalčić et al. (2002) assessed how much of the core-sensitive PKP travel-time data can be explained by mantle structure alone. Using high quality PKP(AB−DF) and PcP−P differential travel-time data, they developed a 3D compressional velocity model of the lowermost mantle and outlined its relevance to understanding anisotropic structure of the Earth's core. A large amount of the PKP(AB−DF) differential travel time data was adequately explained by mantle structure particularly in the lowermost portion, alone. There was no need to invoke IC anisotropy (Figure 4.14a).

South Sandwich Islands Earthquakes

However, in order to explain the most anomalous PKP(BC)−PKIKP differential travel-time residuals (mostly originating from the South Sandwich Islands (SSI) region), alternative explanations were needed, such as the existence of (i) much smaller scale heterogeneity in the lowermost mantle than what can be

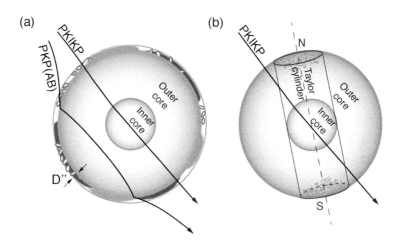

Figure 4.14 Illustration of alternative hypotheses to IC anisotropy. (a) Geometry of ray paths of PKP(AB) and PKIKP through the lowermost part of the mantle and core (solid lines). The lowermost 300 km of the mantle is shown as a layer with multiple heterogeneity scales (D''). (b) Tangent (Taylor) cylinder in the OC and volumetric heterogeneity near polar caps. PKP rays (PKIKP shown here) sample the tangent cylinder depending on source-receiver geometry. Modified from Tkalčić and Kennett (2008).

resolved with the current spatial ray path sampling, or (ii) a combination of heterogeneity and anisotropy in the mantle and core. The latter seems the more attractive and likely scenario. It is difficult to definitively choose between the two because the P-wave ray path coverage of the lowermost mantle is still relatively poor.

Romanowicz et al. (2003) reinvestigated anomalous PKP travel times and contributed a new set of travel times corresponding to events from the southern Atlantic Ocean measured on the Alaskan short period station network. They attempted to fit the anomalous travel times using various IC anisotropy models, various mantle models, and the combination of both. IC anisotropy models predicted the magnitude of the observed residuals but failed to predict the pattern and scatter in the data, whereas mantle heterogeneity models predicted the patterns and scatter in the data better, but failed to predict the magnitude. Combinations of the two worked reasonably well; however, 2.5–3.0 s of the most anomalous travel-time residuals could not be explained by a combination of published models.

A possible explanation for anomalous differential travel times from the SSI earthquakes was the existence of a flat thin slab, or multiple slabs, in the mantle (there is a subduction zone on both SSI and Alaska ends of this path) that were not resolvable by global tomography models. A flat slab of a size comparable to the separation between PKIKP and PKP(BC) ray paths could potentially affect travel times of PKIKP and PKP(BC) waves as to produce the observed anomalous signal seen for this particular source-receiver geometry. An illustration of rays traced through a realistic P-wave velocity model demonstrates the feasibility of this scenario (Figure 4.13).

The alternative explanation was OC structure. Indeed, when polar projections are chosen to display where PKP data pierce the core boundaries, a new pattern emerges: anomalous travel time residuals appear confined to circles about the polar caps. Romanowicz et al. (2003) found that a 0.5–1.0 per cent increase in P-wave speed inside the 'tangent cylinder' (a cylinder tangent to the IC with distinctly different convection characteristics than the rest of the OC and aligned with respect to the Earth's spin axis) could account for the geographical trends in absolute and differential PKP travel-time data (Figure 4.14b). In particular, such structure in the OC can account for the L-shaped pattern of travel-time residuals mentioned previously. While arguably unrealistic due to evidence that the OC is well mixed, this hypothesis prompted further investigation, including a comparison with IC anisotropy models (see Section 4.4.2).

IC Anisotropy Questioned Using Normal Modes

In 1992 Widmer et al. (1991) significantly expanded the database of anomalously split normal modes. In their investigation, they considered various causes for the anomalous splitting of normal modes. For example, they tested a volumetric perturbation, both isotropic and anisotropic, in the IC. They demonstrated the impossibility of reconciling all measured mode splitting with IC anisotropy alone. Some modes have little elastic energy in the IC (typically less than 2 per cent), and some have more than 7 per cent. They were unable to fine-tune anisotropy in the IC to explain the anomalous splitting of both these classes of modes. Increasing the power as to explain the splitting of modes with little energy over-estimated the splitting of highly sensitive IC models. The authors concluded that IC structure was not responsible for the anomalous splitting. Instead, they converged on a solution that had an axisymmetric volumetric perturbation in the OC, similar to the model proposed by Ritzwoller et al. (1986). They recognised that a degree-two perturbation in the OC required implausible physical conditions, namely, density perturbations beyond what was geodynamically sound (Stevenson, 1987).

The large Bolivia earthquake of 1994 provided an unprecedented quality of seismological data, which provoked a number of deep Earth studies and led to new measurements of normal mode splitting. At the same time, supporters of the IC anisotropy hypothesis were investigating potential physical causes. Was it due to frozen growth-related dynamics or convection? Intrigued by this question and armed with new normal mode and travel time data, Romanowicz et al. (1996) demonstrated that a departure from radially symmetric IC anisotropy can in fact provide a reasonable fit to both types of data. While cylindrical anisotropy fit neither travel times nor normal modes convincingly, when the radial symmetry constraint was relaxed, the fit significantly improved. The resulting models of compressional velocity in the IC were elongated in the direction of the rotation axis, with high velocities in the central IC zone, which runs parallel to the rotation axis and peripheral equatorial zones (Figure 4.3f). These results supported low-order convection rather than frozen patterns of IC growth, which would be radially symmetric.

As we saw in Section 4.3.1, Tromp (1995) improved the fit to normal mode and travel time data by adding a depth dependence to anisotropy models. However, one of the biggest problems with this addition was that the most strongly split normal modes are those sampling the shallowest parts of the IC, where, contradictorily, no anisotropy is observed from body waves. This, together with a realisation that the lowermost mantle contribution is a greater source of travel time contamination than previously estimated, led to the reconsideration of OC structure as a possible cause of normal mode splitting. Romanowicz and Bréger (2000) showed that anomalous splitting of normal modes (except for the mode $_3S_2$) could be explained by structure in the OC.

4.4 Complex IC Anisotropy

4.4.1 The Innermost IC (IMIC)?

Research that led to the IMIC discovery/hypothesis

In Section 4.3 we see that the initial motivation behind confirming IC anisotropy and its magnitude gradually refocused during the 1990s on investigating the radial dependence of anisotropy. The difficulty of fitting both normal mode and travel time data eventually led some scientists to reconsider the existence of dominant large-scale anisotropy in the IC. With the rise of broadband instruments and the availability of waveform data in digital formats, many seismologists embraced the opportunity to retrieve seismic waveforms immediately after large earthquakes via the world wide web. However, the data available in the ISC bulletin were considered 'noisy' and less credible than differential travel times carefully measured by waveform cross-correlation. Not everyone gave up on the ISC data, though, for the dataset was large, the volumetric PKIKP sampling of the IC was overwhelmingly better than the newly collected, cross-correlation-based differential travel time data (and thus more comparable with normal mode sampling), and, with the application of proper statistical methods, the effects of noise are significantly minimised.

Ishii et al. (2002a) embraced the ISC data because of the nearly uniform volumetric sampling of the IC by PKIKP ray paths. Despite the relative noisiness of the absolute PKIKP data from the ISC catalogue, their use can be justified when considering the uniformity of normal mode sampling. Ishii et al. (2002a) accompanied ISC data with measurements of mantle and IC sensitive normal mode splitting and available differential travel times of PKP(BC)−PKIKP and PKP(AB)−PKIKP waves to perform an inversion for isotropic mantle structure and transversely isotropic (anisotropic) IC structure. They demonstrated that the absolute PKIKP travel time data and normal mode data can be fit well with a constant IC anisotropy model; however, such a model does not fit the differential travel time data. A trade-off is possible, but requires a more complicated IC anisotropy model or mantle heterogeneity near the CMB. Even radially varying IC anisotropy does not improve the fit. The best models obtained in such a case are those with strong anisotropy near the ICB and weak anisotropy near the Earth's centre, which contradicts various previous findings.

In support of localised heterogeneity in the mantle, rather than complex IC anisotropy, as a likely cause for the complex pattern of travel-time observations, Ishii et al. (2002b) performed a separate study of PKP(BC)−PKIKP and PKP(AB)−PKIKP differential travel time data. In a nutshell, they concluded again that a simple cylindrical anisotropy model of the IC combined with complexity in the mantle satisfies normal mode and absolute and differential PKP travel time observations. They argued that the differences in various IC anisotropy models stem from the paucity of differential travel time measurements and the strong bias that this introduces in the inversion. This directly contradicted all other models of IC anisotropy at the time, including the models that argued for radial and azimuthal variations. Their normal mode data accommodated an isotropic outer layer of the IC if its thickness was less than 150 km, but this contradicted certain travel times. For example, this model was inconsistent with the findings of Ouzounis and Creager (2001) and Song and Xu (2002), who propose a weakly anisotropic outer layer and a strongly anisotropic interior (≥ 8 per cent) to the IC. A thicker isotropic layer directly contradicts normal mode data, whereas a strongly anisotropic interior would produce PKIKP travel time residuals much larger than had been observed.

The solution proposed by Ishii et al. (2002b) was simple, except that in the innermost part of the IC, a satisfactory fit could not be achieved. If the idea of simple IC anisotropy were correct, something peculiar was happening in the innermost 300 km of the IC, something absent from its outer shell. As means of elucidation, let's go back to our example of the physics experiment and data binning.

The Discovery/Hypothesis

The arguments against data binning and averaging were challenged by Ishii et al. (2002b) and Ishii and Dziewoński (2002) and confirmed by Ishii and Dziewoński (2003). We said in Section 4.3.4 that the data binning would only be reasonable if transverse isotropy (cylindrical anisotropy) affected the entire IC. One advantage of having a large dataset is the ability to perform sensitivity tests such as the 'bootstrap technique' to check the robustness of the results. If randomly selected subsets behave the same as the main dataset, then the result is considered robust. Because there is a large number of ISC data, it was possible to put the transverse isotropy hypothesis to this critical test.

In that test, the data were first binned with respect to the epicentral distance at which PKIKP waves were measured, and then with respect to the region of IC sampling (Figure 4.15a). The epicentral distance corresponds to the depth of sampling in the IC (see Table 4.3), and the region of sampling is based on the position of the PKIKP bottoming points. The relationships between the angle ξ (or $\cos^2 \xi$) and the absolute travel time residuals from each of the four groups were very similar (i.e. comparable slope and curvature), except for South Sandwich Islands to Alaska data. Most of the PKIKP absolute travel time residuals for values of $\cos^2 \xi > 0.7$ were consistent with a simple model of anisotropy (black lines in Figure 4.15). This convincingly demonstrated that a quadratic dependence is not an artefact of outliers in a global dataset, but instead a legitimate characteristic of each subset. Therefore, the ISC absolute travel time data were consistent with constant transverse isotropy in the IC, which contradicted a hemispherical model. Any evidence of a hemispherical pattern to IC anisotropy could simply be an artefact of uneven ray path sampling, which could allow localised regions to impose too much weight in the analysis of differential travel time data.

However, in the antipodal distance range of 173°–180°, travel time measurements were not well explained by the best fitting model of constant transverse IC isotropy (Figure 4.15b). When anisotropy in the outermost shell of the IC is corrected for, a large part of signal remains unexplained, but a zonal dependence remains (as indicated by a strong $cos^4\xi$ term; see Figure 4.16). Since all subsets show the same behaviour, the model of transverse isotropy in the innermost 300 km of the IC is robust, but it has different characteristics than the outer IC shell model (Figure 4.3k). In the innermost shell, absolute travel time residuals depend predominantly on the fourth power of the angle ξ. This yields a fast axis in the polar direction, a slow axis at 45°, and a secondary fast axis again along the equatorial plane. Ishii and Dziewoński (2002) obtained less conclusive results for the orientation of the symmetry axis, so resorted to it paralleled the orientation of the outermost 920 km layer (inline with the rotation axis).

If the IMIC exists, it must satisfy two different development regimes in the life of the IC, which is hard to reconcile with the degree one convection in the IC (e.g. Jeanloz and Wenk, 1988; Romanowicz et al., 1996), which would efficiently destroy two distinct IC patterns. Regardless of the geodynamical expectations, the predicted transition at a radius of about 300 km must be confirmed through seismological techniques, which are well suited for probing velocity discontinuities. For example, after Mohorovičić (1910) hypothesised the existence of a discontinuity between the crust and the mantle, he was able to predict its depth based on refracting waves, and subsequent observations of reflecting waves provided confirmation. As for the IMIC, its existence necessitates a considerable change in velocity deep in the IC. This sharp transition would cause either a shadow zone for rays travelling at 45° with respect to the rotation axis (if the P-wave velocity goes from fast to slow at the IMIC boundary) or a triplication for rays travelling parallel to the rotation axis of the Earth (if the P-wave velocity changes from fast to faster). Finally, it had to be confirmed that the IMIC was resolvable assuming the Fresnel zone for 1 Hz waves in the IC is approximately 300 km.

The Search for the IMIC

As we saw in Section 4.4.1, Ishii et al. (2002a,b); Ishii and Dziewoński (2002, 2003) concluded that constant transverse isotropy in the IC is the best overall explanation for the anomalous normal mode and travel time measurements. The existence of an innermost shell within the IC was invoked to explain the most antipodal travel time data. Every time a new hypothesis is proposed in science,

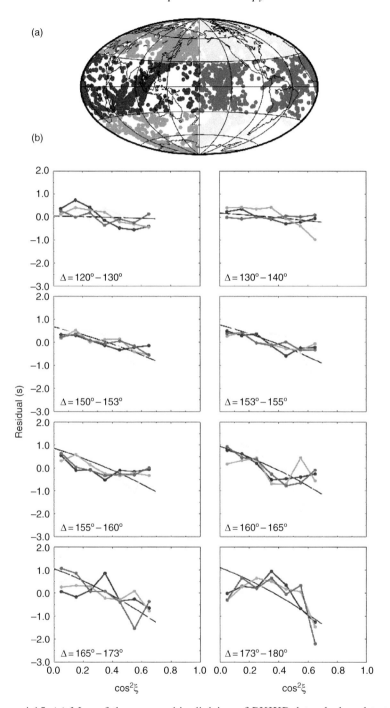

Figure 4.15 (a) Map of the geographic division of PKIKP data, designed to test the robustness of the IC transverse isotropy hypothesis. PKIKP data, based on

it is sustained until it is disproved or contradicted. If a hypothesis 'survives' long enough, it becomes a theory, or, in the case of observational seismology, a confirmed discovery. As for the IMIC, tests are underway and its existence is not yet confirmed, mainly due to the lack of high quality data coverage in the IC, especially by PKIKP ray paths. Thus, it is perhaps a matter of personal preference whether to call the IMIC 'a discovery' or 'a working hypothesis'. Fortunately, there were several lines of research exclusively motivated to test the IMIC hypothesis.

In the pioneering IC studies, normal mode and travel time data were used jointly in inversions, yet the final model of constant IC anisotropy could not explain both sets of observations. The inversion regularisation was another source of issues. Regularisation is a mathematical tool that helps geophysicists overcome difficulties in the inversion, such as lack of information (e.g. data are too few in number or of unknown quality) or overly ambitious model parameterisation (e.g. the basis functions, such as blocks, in tomographic imaging or the Earth layers in 1D profiling are overly discretised). If the blocks in a tomography problem are too small, some will not be sampled by the data at all, making the portion of the model not constrained by the data (a.k.a. the 'null space') large. Here is where the regularisation (damping and smoothing) comes into play and helps to invert the matrix of data. Regularisation helps spread the amount of information more uniformly across the blocks. On the other hand, if the parameterisation is too simple, we can invert with minimal, or no regularisation at all, but we might lose some details. We will discuss the mathematical tools available to counteract this problem in Chapter 5.

The above considerations led Beghein and Trampert (2003) to perform an independent study addressing certain inversion issues. They used normal mode splitting measurements in the inversion and travel times in a forward fashion

---

Caption for Figure 4.15 (cont.) their sampling of the IC, are divided into four regions: eastern polar (|latitude| > 30° ; 0° ≥ longitude ≥ 180°, indicated by green dots), eastern equatorial (|latitude| ≤ 30° ; 0° ≥ longitude ≥ 180°, blue dots), western polar (|latitude| > 30°; 180° ≥ longitude ≥ 360°, yellow dots), and western equatorial (|latitude| ≤ 30°; 180° ≥ longitude ≥ 360°, red dots). Dots on the map are bottoming points of PKIKP waves in the distance range 150°–153°. (b) Comparisons of subsets of PKIKP data at various distance ranges. Data with different colours correspond to quadrants shown in (a). Data are truncated at $cos^2\xi = 0.65$ because they are available for only two equatorial regions. The black curve is the prediction based on the constant anisotropy model from Ishii et al. (2002a). Modified from Ishii et al. (2002b). (A black and white version of this figure will appear in some formats. For the colour version, please refer to the plate section.)

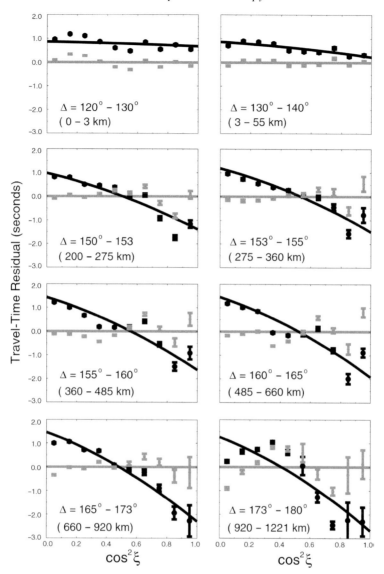

Figure 4.16 Fit to PKIKP travel time residuals at various distance ranges. Observed PKIKP travel time data (black circles with error bars) and residuals after correcting for a constant anisotropy model (grey circles with error bars) for various distance ranges. The values shown below the epicentral distance range correspond to the bottoming depth of the data below the ICB. The prediction based on the bulk IC anisotropy model of Ishii et al. (2002a) is shown as a black curve and the zero line is shown in grey. The standard deviation of the mean is shown as the uncertainty of each averaged datum point. After Ishii and Dziewoński (2002).

to verify their final model. Their normal mode splitting measurements came from several studies of large Bolivia and Kuril Islands earthquakes (Tromp and Zanzerkia, 1995; He and Tromp, 1996; Resovsky and Ritzwoller, 1998), and they assumed global transverse isotropy in the IC with a radial (depth) dependence.

Beghein and Trampert (2003) used five cubic splines as basis functions. If one wants to search for possible distinctive shells in the IC, the innermost one with a radius of, possibly, only 300 km, this parameterisation is a logical choice. The authors recognised that the null space associated with this parameterisation is large. Normal modes are mostly sensitive to the outermost IC, with little sensitivity to the mid depths and even less near the centre of the Earth; this leads to a trade-off. If we want to avoid a large null space, and with it the need to regularise our inversion, we could decrease the number of splines. However, this will inevitably lower our chances to confirm the existence of the IMIC, so the authors decided to keep the five spline parameterisation and introduce damping in their inversion. The linearised inversions with damping produced models with strong anisotropy near the outer region and zero anisotropy towards the centre of the IC Figure 4.17a,b. On an intuitive level, this is because most normal modes do not have any sensitivity to the Earth's centre. Damping suppresses resolution across the entire IC, and since anisotropy in the centre is so weak prior to regularisation, only the outer portion of the core remains anisotropic after damping.

Beghein and Trampert (2003) thus applied a modelling technique that avoids regularisation and more completely samples the model space, which could lead to new models of IC anisotropy free of damping effects. The Neighbourhood Algorithm (NA) (Sambridge, 1999) was chosen for this task. Regardless of the mantle model chosen to correct for the 3D heterogeneous structure (Figure 4.17c), anisotropy near the Earth's centre turned out to be surprisingly large considering the insensitivity of the normal mode data at these extreme depths. Another finding was that most elastic parameters in their model of anisotropy exhibited transitions near a radius of 400 km.

In addition to fitting normal mode data, the model of Beghein and Trampert (2003) predicted differential travel time measurements well for the epicentral distance range 150°–180°. The obtained profile of cylindrical anisotropy had a symmetry axis progressively tilted from about 45° at a radius of 900 km to about 90° at a radius of 600 km. This was compatible with predictions based on the hcp phase of iron. However, the lowermost third of the IC model was incompatible with predictions based on the hcp phase of iron. They speculated that this could evidence a transition between two different phases of iron, both stable at IC conditions.

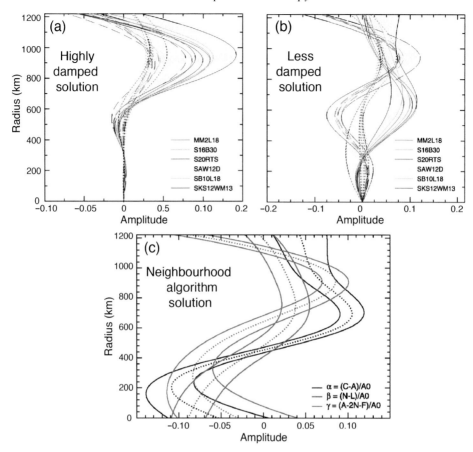

Figure 4.17 Models of anisotropy parameter strength as a function of radius in the IC (0 km = Earth's centre; 1221 km = ICB), resulting from linearised inversion with high (a) and small (b) damping. The solid lines represent P-wave anisotropy; the dotted lines, S-wave anisotropy; and the dashed lines, parameter γ. Different shear wave mantle modes were used to correct the data, yielding the inner-core models in coloured and grey areas. (c) Models resulting from the application of the Neighbourhood Algorithm (NA), where no damping is used. The thin dotted line represents the mean model, and thick surrounding lines correspond to two standard deviations taken from the posterior probability density functions obtained from the NA. Modified from Beghein and Trampert (2003). (A black and white version of this figure will appear in some formats. For the colour version, please refer to the plate section.)

Interestingly, Ishii and Dziewoński (2003) and Beghein and Trampert (2003), although using different methods and travel time data, arrived at relatively simple models of transverse isotropy in the IC with some form of radial dependence. Both their models were void of lateral variations in anisotropy and did not require an isotropic layer in the outermost IC. It seemed that the research on this topic had gone full circle and returned near its starting point, visited some seventeen years earlier (Woodhouse et al., 1986).

Since the aforementioned pioneering studies, there was only a handful of research papers concerning the IMIC. Cormier and Stroujkova (2005) used waveform modelling to search for the IMIC. As forecasted earlier by Ishii and Dziewoński (2003), depending on the direction of sampling of PKIKP waves, the presence of an IMIC would be evidenced by either a triplication or a shadow zone. In addition, at antipodal distances one may expect diffraction around the IMIC. Cormier and Stroujkova (2005) analysed particle motions for weak phases that could be associated with the effects of triplication. However, they found that these phases are not polarised in the same plane as the main IC phases, thus suggesting the root cause resided outside the IC. The IC structural models that agreed with their seismological observations had smooth, larger than 100 km-thick transitions from high to low attenuation between about 400 and 600 km in radius, and uniform attenuation in the centremost shell (Figure 4.31). The transition zone was interpreted as the potential edge of solidification.

Leyton et al. (2005) studied the International Monitoring System (IMS) data, i.e. the precritical reflections of PKiKP waves, for possible discontinuities in the IC. A discontinuity would create a reflection visible in the waveforms. Even after they improved the quality of the signal, they found no evidence for the existence of a widespread radial discontinuity within the upper 450 km of the IC. They argue attenuation reduces the resolving power deeper in the IC and prevents a verdict upon the possible discontinuity in the centremost part of the Earth.

Calvet et al. (2006) demonstrated that the non-uniqueness of the seismological IC anisotropy models could lead to misinterpretations. They showed that at least three different families of IC anisotropy models explained the same data. While the outermost part of the IC (from a 450 km radius to the ICB) appeared cylindrically anisotropic, the remaining, innermost part of the IC seemed to have significantly different anisotropic properties. One family of models consists of a weak anisotropic centre with a slow symmetry axis parallel to the Earth's spin axis (Figure 4.18a). The second family features an almost isotropic IC (Figure 4.18b). The third family is strongly anisotropic, with a fast symmetry axis parallel to the Earth's spin axis (Figure 4.18c). Interestingly, each family of models would have very different consequences for the origin of anisotropy in the IC. The first two could be explained by a degree-one convection pattern (Figure 4.18d), whereas the last one could be explained by a degree-two convection pattern (Figure 4.18e), as supported by Romanowicz et al. (1996). In response, Calvet et al. (2006); Souriau and Calvet (2015) emphasised the importance of accounting for non-uniqueness in geophysical modelling.

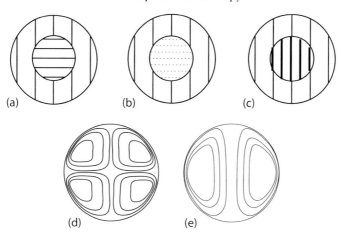

Figure 4.18 The three families of models obtained from an inversion using absolute PKIKP travel time residuals. Model I (a) has a fast symmetry axis parallel to the Earth's spin axis in the outer layer, and a slow symmetry axis in the central part. Model II (b) has a decreasing anisotropy with depth. Model III (c) presents slightly increasing anisotropy with depth but the fast axis remains parallel to the Earth's spin axis. A degree-two convection pattern (d) could explain the anisotropy observed in models I and II. A degree-one convection pattern (e) could explain the anisotropy pattern observed in model III. After Calvet et al. (2006).

4.4.2 IC as a Conglomerate of Anisotropic Domains?

OC Structure

Though OC studies are beyond the scope of this book, we will touch upon them briefly here in relation to the interpretation of IC anisotropy. It seemed as though the real cause for anomalous measurements of normal mode splitting and body wave travel times was playing a hide and seek game with seismologists: at first, it was in the mantle, then the IC, later the OC, and then back to the IC!

We saw in Section 4.3.4 that OC structure as a possible explanation for the anomalous splitting of normal modes received increased attention in the late 1980s (Ritzwoller et al., 1986) and 1990s (Widmer et al., 1992). Some of the normal mode data could be explained by an OC with strong heterogeneity in its tangent cylinder or a concentration of structure in its polar caps (Romanowicz and Bréger, 2000). The former model was supported by the modelling of travel times from the South Pacific and recorded by the Alaskan short period network (Romanowicz et al., 2003). A 0.5–1.0 per cent increase in P-wave speed inside the tangent cylinder could account for the intractable L-shaped pattern of travel-time residuals.. If P-wave heterogeneity cannot be sustained because of the instability of density variations in the OC, P-wave anisotropy in the OC offers a possibility.

Two lines of work, however, demonstrated that the above models are not supported by other types of seismological data. Firstly, Souriau et al. (2003) analysed PKP(BC) waves that sampled the tangent cylinder and avoidic the IC, and multiple P-wave reflections from the lower side of the core mantle boundary that sample the polar caps (S3KS−S2KS). If there is noticeable anisotropic structure in the tangent cylinder, there would be characteristic effects on the travel times of the PKP(BC) waves. If there is heterogeneous structure near the polar caps (or equatorial bulge), there would be a telltale signal in the differential travel times of S3KS and S2KS waves. Neither of the above was observed, however, and they concluded that compared to the IC, the OC's affects on body wave travel times were negligible.

Secondly, Ishii and Dziewoński (2005) analysed the splitting of normal modes sensitive to various parts of the mantle, OC, and IC. By subdividing the dataset, it is possible to understand the influence of OC structure on the data. They found small values for both density and compressional velocity perturbations in the OC (0.06 ± 0.12 per cent for P-wave velocity and -0.06 ± 0.20 per cent for density). Allowing for larger perturbations within the tangent cylinder severely disrupted the fit to OC-sensitive normal modes.

Finally, based on the broadband data from Antarctica, Leykam et al. (2010) conducted a study on the tangent cylinder and its possible structure, concentrating mostly on ray paths sampling the southern part of the tangent cylinder. They found smaller travel time residuals for northern events recorded in Antarctica than southern events observed in Alaska. Because there was no travel time residual anomaly in the southern part of the tangent cylinder, they deemed it unlikely that the phenomenon generating anomalous splitting and travel times occupied the tangent cylinder of the OC. When excluding travel time data from the SSI earthquakes, Leykam et al. (2010) demonstrated that IC cylindrical anisotropy, if it exists, would be weak (≈ 0.7 per cent) and unnecessary to explain the remaining travel time data.

Conglomerate Hypothesis

If the main cause of the anomalous travel times from the SSI earthquakes recorded in Alaska stems from outside the IC, transverse isotropy in the IC would not be required to explain patterns of PKP travel time residuals(see Figure 4.19b). The estimated magnitude of anisotropy when excluding SSI data (≈ 0.7 per cent; Leykam et al., 2010) is small enough to explain the observed travel time residuals with mantle or OC structure, which explains the data scatter better than a simple IC anisotropy model. However, one has to keep in mind that normal mode data require aspherical structure somewhere in the core (see Section 4.2.2).

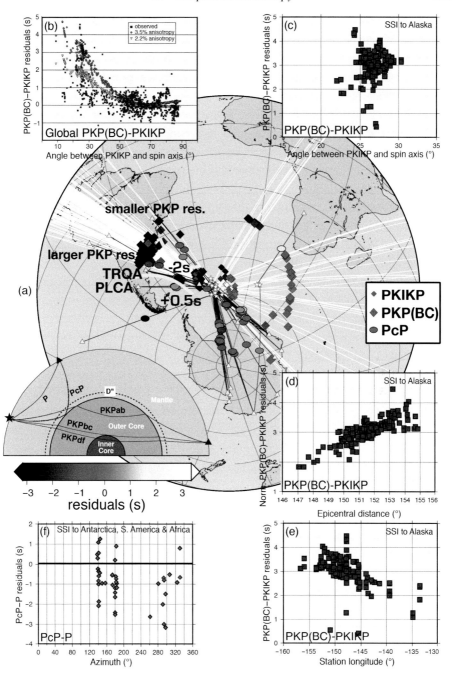

Figure 4.19 (a) Map of the location of SSI earthquakes used in studies of PKP(BC)−PKIKP and PcP−P differential travel times (stars). Reflection points of PcP waves at the CMB are projected to the surface (ellipses) in different colours corresponding to the observed PcP−P differential travel time residuals. Piercing points of PKIKP and PKP(BC) waves at the CMB are projected to the surface

The above developments led to a re-examination of SSI earthquake data (Tkalčić, 2010). The SSI subduction zone can be considered a large factory of earthquakes with significant contribution to global datasets. The earthquakes originate at various depths along the subduction zone. This is compressional environment, so the energy associated with PKP waves is almost always radiated steeply down, near the centre of the radiation pattern lobe, towards the Earth's core. The PcP waves at larger epicentral distances have similar geometrical behaviour and are well excited if the energy leaves the focal sphere near the centre of the radiation pattern lobe. They are poorly excited if the energy leaves the focal sphere near the periphery of the radiation lobe. If PcP−P differential travel times (sensitive to mantle structure only; see Earth cross-section view with PcP and P ray paths in Figure 4.19a) exhibit an azimuthal pattern or dependence on epicentral distance analogous to the PKP travel time data, the cause of the anomalous PKP differential travel times must be in the mantle.

PKP(BC)−PKIKP differential travel time residuals do not depend on the angle ξ (Figure 4.19c), but are linearly dependent on epicentral distance and station longitude (Figure 4.19d,e). The travel time residuals are confined between 0.5 and 4.5 seconds (4-second spread). When we plotted the measured (by cross-correlation) PcP−P travel time residuals for SSI earthquakes and analysed their variations, the scatter was similar to that of the PKP data, where residuals vary between about −3 and 1 seconds: again, a 4-second spread (Figure 4.19f). This new dataset was completely insensitive to core structure, as both P and PcP waves sample only the Earth's mantle. Some of the largest PcP−P residuals occur in the same corridor

---

Caption for Figure 4.19 (cont.) (small and large diamonds) with the corresponding PKIKP−PKP(BC) differential travel time residuals using the same colour scheme. Travel time residuals are relative to the model ak135 (Kennett et al., 1995). PKP and PcP ray paths projected to the surface are shown in white and black lines. Yellow lines indicate a corridor in which some of the largest departure from theoretical predictions in PKIKP−PKP(BC) and PcP-P travel times are observed. GSN stations TRQA (inside the corridor) and PLCA (outside the corridor) are highlighted. A schematic representation of Earth's cross-section and ray paths of seismic phases PKP, PcP, and P waves is shown in inset. (b) Global dataset of PKP(BC)−PKIKP differential travel time residuals as a function of the angle ξ along with the theoretical predictions from 2.2 per cent (yellow inverted triangles) and 3.5 per cent (red diamonds) models of uniform cylindrical anisotropy in the IC. PKP(BC)−PKIKP differential travel time residuals from SSI earthquakes as a function of: (c) the angle ξ, (d) azimuth, and (e) epicentral distance. (f) PcP−P differential travel time residuals from SSI earthquakes as a function of azimuth. Modified from Tkalčić (2010). (A black and white version of this figure will appear in some formats. For the colour version, please refer to the plate section.)

Table 4.4 *Seismological datasets used in the IC studies, and Earth structure/type of anisotropy that can be invoked to explain each dataset.*

	Mantle only	Mantle + IC Cylindrical Anisotropy	Mantle + IC Axial Symm. Anisotropy	Mantle + IC Complex Anisotropy
absolute $PKIKP$	NO	YES	NO	YES
$PKP_{AB} - PKIKP$	YES	YES	YES	YES
$PKP_{BC} - PKIKP$	NO	NO	YES	YES
Normal modes	NO	YES	YES	YES

of azimuths in which the largest PKP(BC)−PKIKP residuals were observed (see the colour scheme for residuals and the projection of piercing points at the CMB in Figure 4.19a). For an epicentral distance of 60°, the difference in P and PcP take-off angles using model ak135 is only 13°, which is a typical take-off angle difference for PKP waves. While it is perhaps a coincidence that the azimuthal patterns of PcP−P and PKP(BC)−PKIKP residuals are similar, it is likely that each of these phases would interact with a slab or its fragments in the mantle and that mantle structure can have a large impact on travel time residuals.

Table 4.4 summarises the types of seismological data explained by a particular model of IC anisotropy or other cause. The table shows four different types of data used to study IC anisotropy and reveals that each model of IC anisotropy fails to explain at least one of these. The mantle structure only model convincingly explains the global PKP(AB)−PKIKP dataset (Bréger et al., 2000; Tkalčić et al., 2002), but other datasets need aspherical structure in the core (see Section 4.2.2). Travel time data from the SSI are most likely affected by mantle structure. Thus, it seems that a combination is required to explain all types of data listed in Table 4.4. These considerations prompted a proposal of a new conceptual model.

Tkalčić (2010) argues that simple cylindrical anisotropy cannot explain the observed scatter in differential travel time residuals (also as demonstrated earlier by Ishii et al., 2002b). But instead of dismissing differential travel times and using PKIKP absolute travel times, it is proposed that the IC is a conglomerate of anisotropic domains (Figure 4.3p). Different domains of anisotropic velocity in the IC, together with contributions from mantle heterogeneity, would give rise to the observed scatter of differential travel time residuals. Whereas normal modes would integrate and smooth out regional IC effects, a model with a net fast axis of anisotropy quasi-parallel to the rotation axis of the Earth would provide a reasonably good fit to the observed splitting of normal modes.

Deuss et al. (2010) used advances in normal mode coupling theory to argue that the coexistence of hemispherical and regional variations of IC anisotropy not only provides a reasonably good fit to the normal mode data, but is required by this dataset. Although different in regards to the prevailing type of anisotropy, the views of Deuss et al. (2010) and Tkalčić (2010) both include a more complex anisotropic velocity distribution in the IC. Thus, near the end of the first decade of this century, IC anisotropy is widely regarded as relatively complex, though occasionally there is argument for a predominantly cylindrical anisotropy which is stronger in the qWH (e.g. Sun and Song, 2008; Irving and Deuss, 2011) (Figure 4.3 m,n,o).

4.4.3 Geodynamical Implications

Mineral Physics Experiments

It is widely accepted in the geophysical community that the hexagonal close packed phase of iron (hcp, also known as *e* phase) is stable at core conditions (e.g. Tateno et al., 2010; Deguen, 2012). It is less certain, however, whether or not this is the only stable phase in the IC. There are two more phases of iron identified from diamond anvil cell experiments: the cubic close packed phase (fcc, also known as *a* phase) associated with high temperatures and the body centre cubic phase (bcc, also known as *g* phase) associated with lower temperatures and pressures. The bcc phase reappears in a narrow stability field just below melting. Although the hcp phase is largely considered the sole survivor in the IC, other observations, such as shock wave experiments, suggest that other stable phases of iron might also exist at IC conditions (e.g. Steinle-Neumann et al., 2001), perhaps due to the coexistence of other chemical elements, such as nickel and oxygen, in the IC (e.g. Vočadlo et al., 2003; Dubrovinsky et al., 2007).

Seismological hypotheses, even when not fully supported by all data, are often given credibility because of the lack of experimentally controlled conditions to suggest otherwise. Geophysical inference (geophysical inversion) is often used as a substitute for direct measurements. IC anisotropy, as characterised by seismologists, has changed in character and strength multiple times. Although the mineral physics community performs measurements in a laboratory environment, the situation has been similar, given that at IC temperature and pressure conditions, the lab equipment and experimental methods are pushed to their limits. As an example, let us consider the measurements of the fast axis of anisotropy for the hcp phase of iron.

Some of the first calculations at low temperatures and high pressure (e.g. Stixrude and Cohen, 1995; Bergman, 1998) suggested that P-wave propagation is faster along the *c*-axis (cylindrical symmetry axis) than along the *a*-axis (in the basal plane, perpendicular to *c*-axis), which agreed with low pressure/low

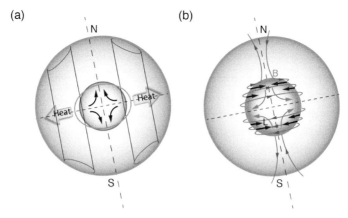

Figure 4.20 Two examples of dynamical models that argue for post-solidification deformation as a predominant cause of anisotropy in the IC. (a) Columnar convection occurs in the OC under the strong influence of the Coriolis force. Convection transports the heat outward in the equatorial region of the IC. The equatorial region of the IC grows faster than the polar regions because of large heat extraction. The inhomogeneous growth results in the isostatic disequilibrium in the IC. The non-hydrostatic stress this creates determines the preferred orientation of iron crystals. Modified from Yoshida et al. (1996). (b) The dynamics of the outer portion of the IC (grey ring) is dominated by the direct effects of the Lorentz force (the Maxwell stress). The dynamics in the deep portions are controlled by the force balance between the pressure gradient and the viscous force, with a constraint imposed by the boundary conditions near the surface of the IC which are dominated by the Maxwell stress. The Lorentz force, caused by the toroidal magnetic field, squeezes inner-core materials towards the rotation axis and causes a flow from strong-field regions to weak-field regions. The flow pattern is sensitive to the geometry of the magnetic field. Black arrows represent the Lorentz force; grey arrows pointing out of the IC core in the equatorial region are flow directions; grey lines coiling around the IC are magnetic field lines. Modified from Karato (1999).

temperature hcp analogs and seismological observations of PKIKP waves. However, the experiments of Mao et al. (1998) and Merkel et al. (2005) suggested a fast direction at an angle between the a- and c-axes. Note that this coincided with the period in which IC anisotropy was becoming increasingly complex. Antonangeli et al. (2006) demonstrated that this intermediate angle was an artefact of iron aggregate texturing in diamond anvil cells. The situation was complicated further with *ab initio* calculations that suggested a reversal at high temperatures to a slow c-axis and a fast a-axis (e.g. Steinle-Neumann et al., 2001; Vočadlo et al., 2009). The most recent diamond anvil studies at IC temperature and pressure, however, all show that the c-axis is slightly faster than the a-axis (e.g. Sha and Cohen, 2010). Unfortunately, the ambiguity of these mineral physics results makes seismological interpretations challenging.

Geodynamical Models

The physical reason behind the alignment of crystals in the IC (which cumulatively would produce anisotropic effects) is not well known, although there are many hypotheses. Some invoke solidification, and some involve post-solidification deformation and/or recrystallisation. The hypotheses that argue for solidification as a cause of anisotropy are: (i) anisotropic paramagnetic susceptibility (Karato, 1993); (ii) the single crystal concept (Stixrude and Cohen, 1995); and (iii) texturing due to directional solidification (Bergman, 1997). The hypotheses that argue for post-solidification deformation as a cause of anisotropy are: (i) IC thermal convection (Jeanloz and Wenk, 1988); (ii) misalignment between the gravitational equipotential and the thermodynamical equilibrium figure of the inner-core field (Yoshida et al., 1996); (iii) radial flow due to Lorenz stresses (Karato, 1999); and (iv) longitudinal flow due to Lorenz stresses (Buffett and Wenk, 2001). Each hypothesis deserves equal attention and consideration when interpreting seismological results.

All of the above hypotheses can be challenged. For example, an attractive explanation for crystal alignment is a post-solidification process. If the IC grows faster in the equatorial direction (because heat flow is fastest in that direction) it would correct its shape in a similar way to glacial isostatic rebound. This is a relatively quick process that leads to imposed strain. However, the strength of this process is highly unlikely sufficient to produce the 3 per cent anisotropy needed to fit the very anomalous travel-time data. Moreover, the timescale of such a process would be very long; longer, in fact, than the age of the Earth.

If IC anisotropy is instead exclusively a consequence of solidification, anisotropy should be stronger in the deeper portion of the IC, which has had more time to solidify. However, such an effect has not been observed seismologically. If deformation exists, and if anisotropy is entirely due to solidification, then post-solidification processes need to be weak enough to preserve the pre-existing fabric.

Laboratory experiments (Bergman, 2003) support the hypothesis that the IC is composed of columnar crystals with cylindrical, not spherical, symmetry. The crystals grow as columns perpendicular to the rotation axis of Earth. If the *c* crystallographic axes correspond to fast axes, anisotropy would depend on their orientation during the solidification process. Consequently, there should be a depth and *c*-axis dependency of anisotropy for PKP waves with equatorial paths.

It is challenging to identify a plausible physical mechanism for a hemispherical (longitudinal) dependence of IC anisotropy. Convection in the OC could be controlled by the mantle (Bloxham and Gubbins, 1987; Sumita and Olson, 1999), which would then leave a signature on the IC texture. However, this would mean that the IC is locked to the mantle (with no differential rotation). Differential rotation aside, the more dire problem with this hypothesis is that the hemispherical pattern supposedly exists to depths of 500 km, which would require structure in the mantle to persist for 500 million years. Mantle convection makes such a long-lived stagnant zone at the base of the mantle implausible.

Candy Wrapper for the IC?

It has been shown through the first principles (Vočadlo et al., 2003) and experimental work (Dubrovinsky et al., 2007) that a body-centred cubic (bcc) form of iron is stable under IC conditions. Belonoshko et al. (2008) argue that iron in a hexagonal close-packed (hcp) form becomes increasingly isotropic with increasing pressure and temperature, and therefore unlikely to cause significant anisotropy. Instead, based on molecular dynamic simulations, they suggest the bcc phase as the more likely candidate to explain the seismological observations. According to Belonoshko et al. (2008), a mixture of isotropic hcp and anisotropic bcc near the Earth's centre can explain the observation of a distinct form of anisotropy in the IMIC (Ishii and Dziewoński, 2002).

The possible coexistence of two iron phases in the IC opens up an exciting avenue towards settling a long-standing seismological controversy that is likely rooted in the uneven spatial sampling of IC sensitive body waves. Mattesini et al. (2010) argue that the presence of anisotropic bcc iron aggregates in the shallower parts of the IC explains the observed scatter in travel time anomalies for the north–south ray paths better than the hcp model. They attribute this hemispherical differentiation to variations in IC heat flow.

In a follow-up paper, Mattesini et al. (2013) show that a mosaic- or conglomerate-like distribution of different iron phases (single crystal hcp and bcc, as well as a cylindrically averaged bcc aggregate) in the IC can explain the complexity of the observed travel times (Figure 4.21). Each crystal orientation of each iron phase can affect differential travel time anomalies quite differently; thus the variety of orientations and phases present in the candy wrapper model successfully predicts the large scatter observed in IC-sensitive differential travel times (Figure 4.22). This model reconciles the seismologically observed hemispherical dichotomy with mineral physics predictions. In the qWH, well-defined and separated bcc and hcp domains form a conglomerate of differently oriented fast crystallographic axes. In the qEH, however, the difference between the hcp and bcc melting temperatures is small, precluding the separation of two different phases of iron.

Towards Further Understanding of IC Anisotropy

We will see in Chapter 6 that modern global seismology is faced with limitations due to the incomplete volumetric sampling of the IC by seismic body waves.

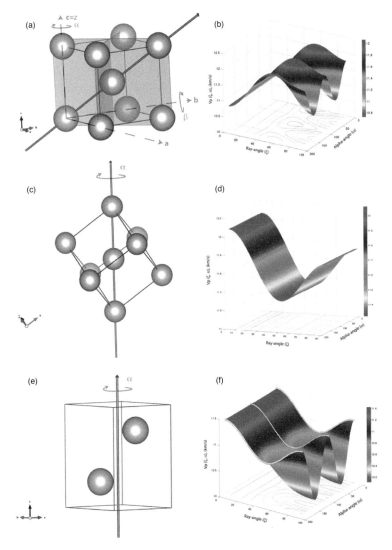

Figure 4.21 (a) *Fe–bcc* unit cell with one of the main diagonal directions [1 1 1] (red vector) representing the fast velocity axis. The two fast velocity planes [1 0 -1] and [1 1 0] at 90.0° with respect to each other are also shown in brown. The α and β angles are clockwise rotations about the *c*- and *b*-axis, respectively. (b) The periodicity of P-wave velocity (v_p) produced by a clockwise rotation about the Earth's spin axis (i.e. from $\alpha = 0°$ to $\alpha = 180°$) for the *Fe–bcc* model, with the fast velocity axis at 54.74° from the main *z*-axis ($\alpha = \beta = 0°$). (c) *Fe–$\overline{bcc}$* unit cell along the vertical crystallographic *c*-axis, aligned parallel to the Earth's rotational *z*-axis. (d) *Fe–$\overline{bcc}$* model, with the fast velocity axis (i.e. the main diagonal) oriented along the vertical crystallographic *c*-axis direction ($\alpha = 45°$ and $\beta = cos^{-1}1/\sqrt{3}°$). Note that there is no velocity modulation for a clockwise rotation about the *c*-axis, thus indicating that the system is purely transversely isotropic. (e) *Fe–hcp* unit cell and its [0 0 1] fast velocity axis direction. (f) v_p of *Fe–hcp* with the fast velocity direction along the vertical Earth's spinning axis ($\alpha = \beta = 0°$) showing the 90° periodicity produced by a clockwise *z*-axis rotation (i.e. from $\alpha = 0°$ to $\alpha = 180°$). After Mattesini et al. (2013). (A black and white version of this figure will appear in some formats. For the colour version, please refer to the plate section.)

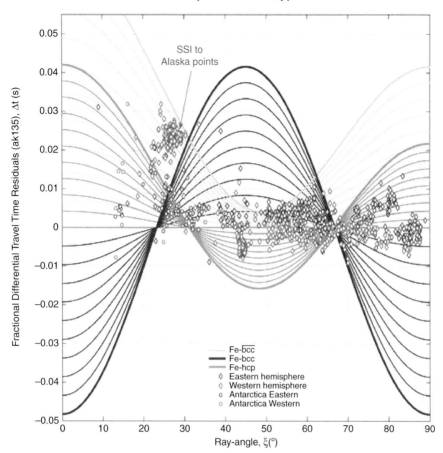

Figure 4.22 'Candy Wrapper' velocity model for PKP(BC)−PKIKP data points for a sampling depth range of 125.8–345.1 km. Data points are shown by blue symbols for the bottoming points in the qEH and by red symbols for the bottoming points in the qWH. The model consists of three sub-models: the conglomerate cubic phase (grey), the bare cubic phase (blue), and the hexagonal close packed iron (green). Note that when omitting the data from SSI region, there is a drastic strength reduction in elastic anisotropy from 3 per cent to 0.7 per cent. After Mattesini et al. (2013). (A black and white version of this figure will appear in some formats. For the colour version, please refer to the plate section.)

There are far fewer quasi-polar paths than quasi-equatorial paths in the IC (for the definition see Chapter 4), which is a leading obstacle to further progress on understanding IC anisotropy. Several ideas are proposed in Chapter 6 for improving the volumetric sampling of IC-sensitive ray paths with the goal of progressing our understanding of the IC. A good example is seismic interferometry, which is one of the rapidly emerging methods in deep Earth seismology.

Seismic interferometry has the potential to revolutionise our understanding of the IC. In a recent paper, Wang et al. (2015) analysed the autocorrelation of earthquake codas and found that the differential travel times of $PKIKP^2$ and $PKIIKP^2$ body waves sensitive to IMIC structure vary by up to 10 seconds (for the definition of these waves, see Section 6.2.4). Based on these observations, Wang et al. (2015) proposed that the IMIC has a fast axis near the equatorial plane through Central America and southeast Asia, contrary to the north–south orientation of the fast axis of anisotropy in the outer IC. However, in a follow-up paper, Romanowicz et al. (2016) argued against a fast axis of anisotropy aligned with a quasi-equatorial direction. Based on the observations of antipodal PKIKP travel times they obtained a significantly better fit to the data using a fast axis of anisotropy aligned with Earth's rotation axis. In a recent investigation of seismic anisotropy Lincot et al. (2015) ruled out bcc structure as plausible in the IC. They argued for hcp structure (Lincot et al., 2016), although Romanowicz et al. (2016) argued that it is not possible to distinguish between hcp and bcc structures in the IC.

These recent works indicate that a long-standing seismological controversy on the nature and origin of IC anisotropy is still alive.

5

Inner Core Rotational Dynamics

"This is not just a grand curiosity; it has implications for the dynamics of the whole planet and for the generation of the Earth's magnetic field."

Comment on "Seismological evidence for differential rotation of the Earth's inner core" published in the same issues by Song and Richards (1996).
Kathy Whaler and Richard Holme, *Nature* (Whaler and Holme, 1996)

"Right now, the differential rotation of the inner core is not yet firmly established."
Annie Souriau, *'Is the rotation real?', Science* (Souriau, 1998b)

"Observations from our laboratory and the independent confirmations of travel time changes using our original method, which document a pattern of change that implies differential rotation of the inner core, should not be obscured by different or negative results from methods that are less capable." *In response to Souriau (1998b).*
Paul G. Richards et al., *Science* (Richards et al., 1998)

"Thus, there is no undeniable demonstration of the existence of inner core rotation. But there is also no undeniable demonstration of the absence of rotation."
Annie Souriau, *Science* (Souriau, 1998a)
In response to Richards et al. (1998).

5.1 Introduction to IC Rotation

The above remarks clearly illustrate the controversy associated with the detection of differential rotation of the IC. It is a fascinating topic that, one could say, gradually overtook, or at least rivalled in its intensity, the seismological research on IC anisotropy.

It is easy to understand the captivation of the seismologists who strove to detect this phenomenon. The changing geomagnetic field observed at the Earth's surface mostly reflects the processes at the top of the liquid OC. However, due to the geomagnetic field's relationship with IC dynamics, seismological observations of IC rotation could invaluably constrain the dynamics of the geomagnetic field at the deeper, inward-facing surface of the OC. Seismologists not only posed the question, "How do we measure IC rotation?", but also, "Can we even detect it?". Doubt was rooted in the pioneering seismological papers on IC rotation. Perhaps the most relevant question was, "What is the magnitude and direction of differential rotation?". This was important because it could indirectly provide relevant information about the IC, such as its viscosity, gravitational relationship with the lowermost mantle, and even its age.

5.2 The Detection

In Halley's Hollow Earth model (Halley, 1686), 'the inner globe' in the Earth's centre rotates differentially with respect to the rest of the planet (Figure 1.1a). The fact that similar phenomena occur in nature may explain how the idea of differential rotation of the Earth's internal shells goes as far back as the seventeenth century. For example, the planets of our solar system exhibit differential rotation around the Sun. Galaxies and protostars also show differential rotation. In stars' interiors, convection constantly redistributes the angular momentum, thus causing angular velocities to be radially dependant. Differential rotation also occurs on stellar surfaces; for example, on the surface of the Sun, plasma rotates faster at the equatorial than at the polar belts.

The NASA *Hinode* Solar Mission satellite recently recorded data that convinced astronomers that the Sun's magnetic field is much more complex than previously believed. The differential rotation at the surface twists and tangles the magnetic field lines of the Sun, which leads to magnetic field reversals. Solar super-tornadoes plough through the Sun's atmospheric layers in a form of hot gas and twisted magnetic lines powered by the nuclear reactions in the solar core (Wedemeyer-Bohm et al., 2012).

It is not too difficult to imagine that the Earth's magnetic field must be spectacularly complex too. Observations of the magnetic field at the Earth's surface, usually presented in terms of spherical harmonics, offer just a glimpse of the actual complexity of the Earth's magnetic field. The differential rotation of the Earth's IC emerges in geodynamo simulations (Gubbins, 1981), where its strength and direction are sensitive to the imposed viscous boundary conditions at the ICB (e.g. Glatzmaier and Roberts, 1996; Kuang and Bloxham, 1997) and the balance between the gravitational and electromagnetic torques (Aurnou and Olson, 2000).

5.2.1 Seismological Methods for IC Rotation Detection

If we were seismologists in the 1990s motivated by recent geodynamical simulations of a differentially rotating IC (Glatzmaier and Roberts, 1996), how could we detect such IC motion? How could we constrain through observation the predictions based on numerical simulations? Could we quickly devise and efficiently apply a suitable new method to existing data?

Fortunately, we would soon discover an ace up our sleeve. Given we already would have access to travel times of many seismic phases recorded globally, the missing ingredient needed to solve the above problem is sensitivity to not just the three spatial dimensions, but also to a fourth dimension, time. Detecting IC rotation requires repeated measurements of travel times of IC-sensitive phases. This is comparable to performing repetitive measurements of temperature and other meteorological parameters to study climate change. Repeated travel time measurements along the same IC-sampling path at different points in time would bring us closer to detecting possible IC rotation. Fortunately for us, earthquakes and explosions often reoccur in approximately the same location and stations generally operate for prolonged time periods, meaning that we have similar paths to probe the IC at different points in time. But how could we decouple the effects of structure in Earth's upper shells and assess the motion of its innermost shell?

We can define differential travel time Δt as

$$\delta t = [PKP(BC) - PKIKP]_{meas.}, \tag{5.1}$$

where *meas.* denotes the measured (observed) difference in onset times of PKP(BC) and PKIKP phases. Note that this is the first term of Equation 1 in Box 2.1. In the above equation, the total PKIKP travel time is compared with that of the neighbouring PKP(BC) ray, which has a similar path through the mantle, but stays inside the liquid OC. The travel time of each phases is a summation of time spent in the crust, mantle, the OC, and the IC. However, it can be assumed that structural effects of the crust and mantle cancel out due to the close proximity of the two rays and that there is no significant temporal change in OC structure on the timescale defined by the measurements. The validity of this assumption can be easily tested on a dataset of PKP waveforms from a given seismogenic zone recorded on a single seismographic station over a given time interval. If the onset of the IC-sensitive phase PKIKP varies while the onsets of the PKP(BC) phase is invariant, IC structure must have changed. Thus, any significant variation in δt for earthquakes from the same region recorded at a single station indicate temporal variation in IC structure. Since the IC does not undergo vigorous convection, a remaining possibility is that the entire IC is rotating as a rigid body.

A potential method of IC rotation detection is to use a known patch of heterogeneity in the IC that differentially rotates relative to a source-receiver path fixed in the mantle reference frame. Of course, this would not work if IC structure is spherically symmetric and the spin axes of the IC and the mantle are coincident. Luckily, this appeared not to be the case, as the existence of IC heterogeneity is supported by a number of studies (e.g. Shearer, 1994; Tanaka and Hamaguchi, 1997). Thus, a patch of heterogeneity with known characteristics can be used as a marker to measure the speed of rotation.

IC Rotation Detection Methods Using Anisotropy

Although the 'marker' method sounds promising, this was not the way that IC differential rotation was first detected. Instead, IC cylindrical anisotropy provided the crucial framework for proving, or disproving, IC rotation. This is not surprising considering that IC rotation work coincided with the peak in IC anisotropy work. It was also a practical choice, because IC anisotropy was better understood than IC heterogeneity. What follows is a simple explanation of how the differential rotation detection method works.

There are two ways to detect IC rotation by utilising cylindrical anisotropy. The first way is to use a similar ray path, such as required by the method that uses 3D heterogeneity as a marker. If the symmetry axis of cylindrical anisotropy is aligned with regard to the Earth's spin axis, then there is no hope of detection. Cylindrical symmetry means that the material properties are invariant with rotation about the symmetry axis, so if the IC rotates faster (or slower) than the mantle, the PKIKP ray would always sample the IC at the same angle ξ with regard to the Earth's spin axis. However, several studies demonstrated that the fast axis of anisotropy is tilted with regard to the Earth's spin axis, so when the fast axis of anisotropy rotates around the Earth's spin axis, and the angle ξ between the PKIKP path and the fast axis changes in time (Figure 5.1a). This change is proportional to the observed change in PKIKP onset times, thus empowering us with a tool to measure IC differential rotation.

The second way of employing anisotropy to detect IC rotation does not rely on a single source-receiver path, but on a large dataset of PKIKP wave travel times. Assuming that cylindrical anisotropy is stationary with a fast axis tilted relative to the Earth's spin axis, we can determine the direction of the symmetry axis of anisotropy by fitting a 'ring' to the zonal distribution of travel times when projected onto the globe (Figure 5.1b). If we repeat this procedure and determine the direction of the fast axis in space at consecutive time intervals, we can construct a time history of the rotation (and 'the wobble' if there is change in both longitude and latitude of the axis).

(a)

(b)

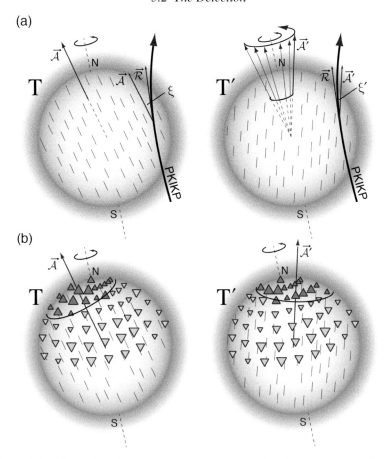

Figure 5.1 Schematic of two seismological methods for detecting IC rotation that involve IC anisotropy with its fast axis tilted with respect to the rotation axis of Earth. Snapshots of IC at earlier time (T) and later time (T′) are shown on the left and right, respectively. The IC is shown by the central sphere embedded in the liquid OC. Anisotropy orientation is shown by thin lines throughout the bulk of the IC and by vector A or A′. (a) The detection method of Song and Richards (1996) that utilises a fixed or quasi-fixed source-receiver path. (b) The detection method of Su et al. (1996) that utilises PKIKP data bins.

IC Rotation Detection Methods Using Heterogeneity

What if, however, the fast axis of IC anisotropy is not tilted with respect to the spin axis of the Earth, or if the tilt cannot be reliably detected? We would need to abandon methods that rely on IC anisotropy and instead utilise IC heterogeneity, as mentioned at the beginning of Section 5.2.1; if we have adequate spatial coverage along a given ray path in the IC and a well-constrained lateral velocity gradient in the IC, we can determine the magnitude of IC rotation by using this structure as a marker.

(a)

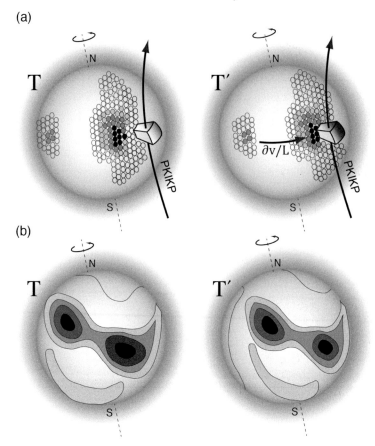

(b)

Figure 5.2 Schematic of two seismological methods for detecting IC rotation that do not involve IC anisotropy. The first one utilises PKIKP travel times along a fixed source-receiver path. Snapshots of the IC at earlier time (T) and later time (T′) are shown on the left and right, respectively. The IC is shown by the central sphere embedded in the liquid OC. (a) The detection method of Creager (1997) that utilises a fixed or quasi-fixed source-receiver path. Lateral variations in IC velocity structure are shown by grains of different shades of grey. As the IC rotates with respect to the mantle, velocity structure in the volume sampled by the fixed ray path changes (low to high velocity change is shown here by a transition from white to black). (b) The detection method that utilises normal mode splitting functions, as used by Sharrock and Woodhouse (1998) and Laske and Masters (1999). Splitting functions derived from a particular mode of vibration are sensitive to radial velocity structure. If the IC rotates as a rigid body, then the shift observed in splitting functions over time can be used to determine the IC rotation rate.

The 'marker' method is illustrated schematically in Figures 5.2a and 5.3. When a ray path is fixed in the mantle reference frame, it will sample different velocities along a velocity gradient in the IC as the IC rotates with respect to the mantle. We can estimate the speed of rotation by studying the effect of a known IC

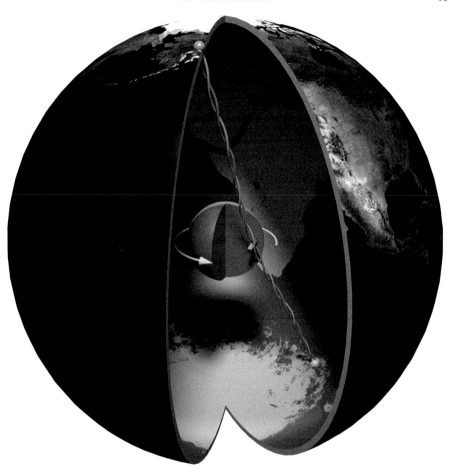

Figure 5.3 Illustration of the rotational dynamics of the IC. Marked are the South Sandwich Islands archipelago (the location of earthquake doublets) (yellow ball) and the COLA seismological station near Fairbanks, Alaska (green ball). The IC is shown at the Earth's centre. The PKIKP ray paths for an earthquake doublet are shown in blue and red. They traverse the western hemisphere of the Earth's IC (for a definition of hemispheres, see Section 3.8.1) through a well-studied portion featuring a linear gradient in isotropic velocity, which is shown by a small patch of varying colour. As the IC rotates eastwardly with respect to the mantle (indicated by the yellow arrow), a fixed source-receiver path will sample the same mantle and OC (not shown) structure, but slightly different IC structure due to the velocity gradient. If velocity gradient is used as a 'marker', it is possible to estimate the differential rate of the IC rotation from the measured time differences between the onsets of PKIKP and PKP(BC) waves (see the text). (A black and white version of this figure will appear in some formats. For the colour version, please refer to the plate section.)

velocity gradient on travel times from SSI earthquakes recorded in Alaska. Combining measurements of Δt (Equation 5.1) with independent constraints on the IC heterogeneity sampled by the ray paths enables estimation of the differential rotation rate of the IC. A given portion of the IC will be well sampled by rays following a quasi-fixed source-receiver path from the SSI to Alaska because of the abundance of earthquakes in the SSI region and the continuous and prolonged recording capabilities in Alaska (Figure 5.2a).

Equation 5.1 defines the differential travel times, δt, which are sensitive to IC structure and can be obtained from a seismogram. Theoretical δt can be expressed as

$$\delta t = \int_{IC} \frac{\Delta v}{v_0} dt + \delta t_{outside\ IC} = \delta v(\phi, \Delta) \int_{IC} dt + \delta t_{outside\ IC}, \qquad (5.2)$$

where Δv is the velocity perturbation relative to v_0, the reference model velocity. $\delta t_{outside\ IC}$ is the contribution to the differential travel time from the crust, mantle, and outer core. To a first-order approximation, this contribution can be neglected for PKIKP and PKP(BC) ray paths; alternatively, it can be more accurately determined and considered. The travel time integral, $\int_{IC} dt$, is the total travel time of PKIKP through the IC for a given reference model, which is readily calculated (see Table 4.3). The velocity perturbation, δv, is averaged along the ray path through a constant, linear velocity gradient in the IC. δv can be expanded to a first-order Taylor series with respect to azimuth of the PKIKP ray turning point, ϕ, and the corresponding epicentral distance, Δ. Following Creager (1997), we obtain

$$\delta v(\phi, \Delta) = \delta v_0 + \frac{\partial v}{\partial \phi}(\phi - \phi_0) - \frac{\partial v}{\partial \phi}\frac{\alpha \Delta T}{\gamma} + \frac{\partial v}{\partial \Delta}(\Delta - \Delta_0), \qquad (5.3)$$

where ΔT is the time lapse between the earliest and latest measurements (assuming that the IC rotates with respect to the mantle at a constant rate for the entire time), and v_0, ϕ_0, and Δ_0 are the values of the velocity anomaly, azimuth, and distance associated with the bottoming point within the heterogeneous subvolume of the IC of the ray chosen as the reference ray. α is the rotation rate of the IC with respect to the mantle, and γ is a dimensionless factor relating the change in distance along the sampling profile in the IC to the change in longitude. For paths between the SSI region and College station, γ is estimated to be 1.1.

The second term in Equation 5.3 represents the contribution from the velocity gradient along ray azimuth, ϕ. Given a constant velocity gradient, the further we move from the reference azimuth, ϕ_0, the greater the change in velocity. The third term accounts for the IC differential rotation and is proportional to the product of the rotation rate, α, and the IC rotation time between the earliest and latest earthquake, ΔT. The rotation term is negative because the IC velocity decreases westwardly with increasing azimuths (from $\approx$ 295 to 325° for the ray paths from SSI to Alaska, as in Figure 5.4). Under this convention, an eastward rotation of the IC with respect to the mantle is positive. The last term denotes velocity change with depth (the larger the epicentral distance Δ, the deeper the rays sample), but because this depth variation is very small compared to the lateral variation, it has minimal impact here.

Unknown parameters such as the crustal and mantle corrections expressed in Equation 5.2 and the rotation rate, α, can be determined through inversion. When we have repeated δt measurements at different points in time from the same ray path through the IC (such as occurs with earthquake multiplets), the second term is zero, as there is no change in reference azimuth.

Instead of expanding the velocity perturbation in a Taylor series with respect to ray azimuth, we can conveniently reference other parameters, such as the distance along the IC sampling profile, L, where the velocity gradient is assumed to be the steepest. The resulting expression for δv is of the same form as Equation 5.3, but the third term contains $\frac{\partial v}{\partial L}$ instead of $\frac{\partial v}{\partial \phi}$.

For the case of an earthquake doublet, we can use Equations 5.2 and 5.3 to write an expression for velocity perturbation:

$$\delta v(L) = \frac{\delta t}{\int_{IC} dt} = -\frac{\partial v}{\partial L} \frac{\alpha \Delta T}{\gamma}. \tag{5.4}$$

We can then express the IC differential rotation rate, α, as

$$\alpha = -\gamma \frac{\delta t}{\int_{IC} dt \, \Delta T} \frac{1}{\frac{\partial v}{\partial L}}, \tag{5.5}$$

where $\frac{\delta t}{\Delta T}$ is the differential travel time slope that can be determined empirically from one or more doublets. The measured differential travel time, δt, is normalised by the total time the ray spends in the IC, and is expressed as a percentage. If the time lapse, ΔT, is expressed in years, the dimension of the IC differential rotation rate, α, is $\frac{\circ}{yr}$.

The velocity perturbation expansion can include additional parameters, such as IC anisotropy (e.g. Song, 2000a), which can be determined through inversion. From Equation 5.5 and the above-described sign convention, if the IC spins eastward relative to the mantle, PKIKP waves of more recent earthquakes will traverse the IC more quickly because they will pass through increasingly faster material. This method (Creager, 1997) has been used in a number of seismological studies of IC rotation, as several locations in the IC have been identified as heterogeneous and have velocity gradients that make suitable 'markers'. In Section 5.4 we will investigate whether or not it is safe to model the IC velocity gradient as

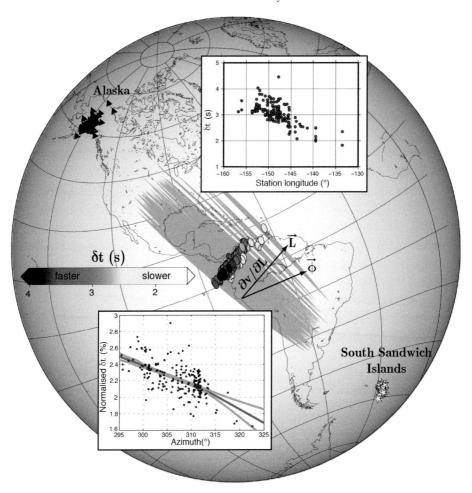

Figure 5.4 The location of the South Sandwich Islands earthquakes (yellow stars), stations in Alaska (triangles), and bottoming points of PKIKP waves (ellipses) are projected to the surface. Bottoming points of PKIKP waves are shown in colours corresponding to the travel time residuals, δt, defined in Chapter 2 in the box on PKP travel time residuals (Box 2.1, Equation 1). Great circle path projections of PKIKP waves in the IC are shown by grey traces. The direction of the steepest gradient in velocity, $\frac{\partial v}{\partial L}$, is marked by $\vec{L}$, while $\vec{\phi}$ marks the changing longitude. The diagram at the top presents the time residuals, δt, as a function of the recording station longitude. The residuals normalised by the time the PKIKP ray spends in the IC are shown as a function of azimuth (measured from the source to the station) in the diagram at the bottom. The expected regression model obtained by averaging all resulting models in a transdimensional Bayesian inversion (see Appendix D) is displayed by the red line. Grey lines represent one standard deviation of the ensemble at each azimuth. Adopted from Tkalčić et al. (2013a). (A black and white version of this figure will appear in some formats. For the colour version, please refer to the plate section.)

perfectly linear using transdimensional Bayesian inversion (Tkalčić et al., 2013a) (see Appendix D).

The phenomenon of IC rotation also attracted seismologists specialising in the long period data from Earth's free oscillations. The splitting functions of normal modes are a visual representation of how each normal mode 'sees' Earth structure as a radial average. The sensitivity of each mode is uniquely distributed along Earth's radius. This is useful because modes with increased sensitivity to the IC can be compared with the modes sensitive only to mantle structure. We can compare the splitting functions of the same mode at two different times (similar to the fixed ray path method) by looking for shifts in their harmonic degree expansion patterns to detect changes in Earth structure (Figure 5.2b). If the splitting functions of mantle sensitive modes do not change, but those of the IC sensitive modes do, we have a strong observational evidence for time-varying structure in the IC. This class of studies sparked an interesting interaction between the body wave and normal mode communities, as is commonly observed in the modern history of global seismology.

Other methods were also invented to substantiate the rotation of the IC. Most notably, there were several seismological studies that used short-period seismological data and the principles of geometrical scattering to detect and measure the rotation rate.

5.2.2 Detection of Simple (Steady) Rotation

PKIKP waveforms observed in the late 1980s at the Warramunga Station in Australia were the first to be explained by temporal changes in IC properties (Souriau, 1989). At the time it was known from the work of Gubbins (1981) that differential rotation of the IC is a physically plausible outcome that emerges in geodynamo simulations. Souriau's observations were nonetheless received with scepticism by the seismological community (A. Souriau, personal communication).

By the time the first journal paper was published on the detection of IC rotation several years after Souriau's conference report (Souriau, 1989), there was ample data from the SSI observed in Alaska accumulated through previous studies of IC anisotropy. This dataset was perfect. On the southern side of the quasi-fixed ray paths, we have the repeating SSI earthquakes. And on the northern side, we have a seismological station at College near Fairbanks, Alaska, that has been recording continuously for several decades. Though there are several other ray path geometries between seismically active subduction zones and seismological stations with a long historical record of operation, not many of them are separated by an angular distance at which PKP waves are observed.

Song and Richards (1996) analysed several geometries, but focused on systematic variations in travel times of PKP waves observed for the SSI earthquakes recorded in Alaska. They reported an approximate 0.3 s increase in PKP(BC)−PKIKP differential travel times over the period of 28 years. They interpreted this increase as a signal that the IC was spinning faster than the mantle, which was precisely in line with geodynamical predictions. The main assumption in their study was that the fast axis of cylindrical anisotropy is tilted with respect to the Earth's rotation axis as discussed in Section 5.2.1. A quasi-fixed source-receiver ray path can be thought of as a reference frame relative to which the fast axis of cylindrical anisotropy moves in time (Figure 5.1a). If the angle between the PKIKP ray path in the IC and the fast axis of anisotropy decreases with time, PKIKP waves will arrive earlier, whilst the part of the waveform traversing the OC, the PKP(BC) phase, will remain unaffected.

A grid search can be performed to determine the location of the pole of the fast axis of anisotropy. As the trial location is varied, the misfit between the observed and predicted differential travel times is calculated. The optimal pole location will yield the least misfit, and hence, the greatest variance reduction. The maximum variance reduction obtained this way is actually not significantly better than the variance reduction based on a stationary IC with cylindrical anisotropy (without the tilt of its symmetry axis). However, the optimal location of the fast axis of the symmetry pole based on the grid search method agrees well with the findings of Su and Dziewonski (1995). Based on the above considerations, Song and Richards (1996) declared the rate of IC rotation to be approximately 1°/yr faster than that of the mantle and crust.

In December of the same year, Su et al. (1996) published a paper that utilised a large dataset of PKIKP travel times. As we saw in Section 5.2.1, if we determine the location of the pole of the fast axis of anisotropy by fitting a zonal pattern to the measured travel time data at two different points in time, we can determine the speed of the rotation of the fast axis of anisotropy, and therefore, the speed of IC rotation (Figure 5.1a). When this procedure is repeated for multiple time intervals, a more detailed history of the IC motion emerges, which includes possible wobbles when there is a change in latitude of the pole of fast axis of anisotropy. Su et al. (1996) performed such an experiment using the same 29 years of data from the ISC bulletins that Su and Dziewonski (1995) used to obtain the direction of the fast axis of anisotropy. They divided the 29 years into six time intervals, each five years long, with an overlap of one year for the time intervals 1984–1988 and 1988–1992. They obtained an IC eastward rotation of 3°/yr faster than the mantle without any significant change in latitude (no wobble). They then generalised their method using spherical harmonics, and instead of monitoring only a general drift of the fast axis of anisotropy, they considered the entire velocity field as defined by travel time residuals. They confirmed their initial result. They also found that in the time interval 1969–1973, there was an anomalously large change in the rotation rate, which they hypothesised to be related to the occurrence of a geomagnetic jerk during 1969–1970.

The scientific community soon witnessed yet another development in detecting a differential rotation of the IC. A new method, first devised by Creager (1997), abandoned the need for uniform cylindrical anisotropy, and is fully explained in Section 5.2.1. The obtained IC rotation was approximately 0.3°/yr faster than the mantle (Creager, 1997), and although much lower than previous estimates, it confirmed the hypothesis of a 'super-rotation' of the IC.

5.2.3 First Challenges

As we saw in Section 5.2, three independent studies exploited different ways to analyse travel time data, and resulted in a unanimous conclusion that the IC spins faster than the mantle. The reported rates differed in magnitude, but each was assumed constant. Souriau et al. (1997) and Souriau (1998b), however, disputed the validity of these detection methods, prompting a series of papers that argued intensively for and against the ability to detect IC rotation.

Souriau et al. (1997) first showed that the direction of the fast axis of anisotropy cannot be uniquely and reliably determined from the available data using geometrical arguments. Because the unsampled volumes of the IC are large, the spherical harmonics analysis such as performed by Su et al. (1996) will tend to accentuate the strongest travel time residuals and create a false tilt of the fast axis of anisotropy (see Figure 5.5). Because the five-year time intervals are not long enough to contain a uniform distribution of travel time residuals across the globe, the estimates of the position of the direction of the fast anisotropy axis, and therefore the estimate of the IC rotation rate, are biased.

Souriau et al. (1997) warned of another possible bias associated with the observation of travel time changes along the single path connecting the SSI and Alaska (as used in Song and Richards, 1996). The issue revolved around instrumentation changes: short period instruments that were operating at earlier times were later replaced by broadband instruments. Song and Richards (1996) addressed this change by correcting the broadband data to short-period data (by applying the corresponding instrument response). However, a systematic bias was introduced through the selection of larger magnitude events in earlier years, when the signal-to-noise ratio was not as good as in more recent years due to less optimal instrumentation. Souriau et al. (1997) showed that when the dataset is modified such that the number of large events is balanced over time, the variation of travel time residuals is reduced, thus decreasing the estimated IC rotation rate. This effect is due to the data measurement process. When differential travel times are measured through cross-correlation (as done by Song and Richards, 1996), the

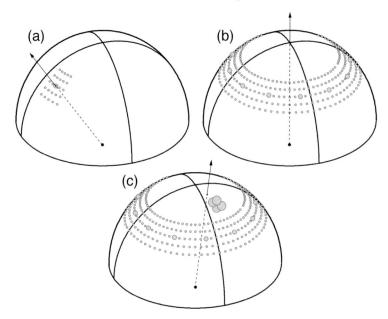

Figure 5.5 Example of bias on the calculated position of the symmetry axis (arrow) resulting from an uneven geographic sampling of synthetic travel time residuals that illustrate bias from real data. The travel time residuals are shown on the northern hemisphere by open circles of a size proportional to their value (the smallest circle = 0.3 s; the medium circle = 0.6 s; the largest circle = 1.2 s). (a) a patch of travel time residuals of 0.3 s with a residual of 0.6 s in the centre. (b) same patch as in (a), but now forming a ring (longitudinal repetition). (c) same as (b) but with addition of four points of 1.2 s. Note how the resulting position of the symmetry axis shifts towards the added group of residuals. Adopted from Souriau et al. (1997).

separation between two phases will decrease for large (and, in this case, older) events when the PKIKP pulse is broadened due to broad frequency content. Souriau (1998b) summarised problems with measurements and assumptions that eventually might lead to artefacts and erroneous inferences about the IC rotation rate. The chief concern was not whether or not the IC rotation existed, but if it could be detected given the seismological data and techniques on hand. Even the small differential rotation rate estimated by Creager (1997) was interpreted with caution and as a potential sign that the rotation of IC might actually be synchronous with the mantle due to the significant uncertainties associated with earthquake location and the likely contamination of travel times by short-scale inhomogeneities in the crust and mantle. An analysis of the polar path between the Novaya Zemlya nuclear explosions and the Dumont d'Urville (DRV) station in Antarctica yielded temporally stable travel time residuals (Souriau, 1998c). This precluded a differential IC rotation rate of as high as 3°/yr, but allowed for the existence of a small differential rotation rate given the uncertainties associated with the prediction estimates.

Soon after the first exchange of arguments about IC rotation (see quotations presented in the introduction to this chapter), Souriau and Poupinet (2000) devised an ingenious method of testing the IC differential rotation hypothesis with a global dataset of PKP wave travel times from the ISC catalogue and hand-picked measurements. They performed a number of forward experiments, each time correcting for mantle structure and assuming a constant IC rotation rate (either negative or positive) with respect to the mantle. They divided PKIKP sampling of the IC into bins. For each rotation rate, they predicted the longitude at which the PKIKP waves would sample the IC. For a given ray path geometry, PKIKP waves will sample different longitudes of the IC depending on the rotation rate, though the latitude would remain the same if the axes of IC rotation and mantle rotation are aligned. Each rotation rate was accompanied by a general trend for the travel time residuals in each bin. They then compared the predicted trends with the observed trends for each bin. Based on that, they obtained that a null rotation (differential rotation of 0.0°/yr) gives the best fit, because it had the smallest absolute mean standard deviation over all bins.

Souriau and Poupinet (2000) identified two problems with a steady rotation: 1) incompatibility with the observed length of day (LOD) fluctuations; 2) incompatibility with gravitational coupling between the IC and mantle.

The first problem can be overcome if the electromagnetic torque is strong enough to compensate for the gravitational torque. The second problem surrounding the coexistence of gravitational coupling and steady super-rotation is only resolvable if the IC can isostatically adjust to the mantle's gravity through viscous relaxation (Buffett, 1997). In this case, the external part of the IC will be constantly perturbed, preventing large scale heterogeneities to develop and become 'markers' for detecting differential rotation.

The debate illustrated via the quotations at the beginning of this section continued into the first half of the twenty-first century (see Table 5.1 for a summary of the studies). On one side of the argument, Song and Li (2000) considered a new path from Alaska to the South Pole and confirmed a 0.6°/yr super-rotation rate using the approach introduced by Creager (1997), thus abandoning the need for a tilt to the IC anisotropy axis as the main tool of detection. They inverted jointly for the rotation rate and IC structure, and found no correlation between earthquake magnitudes and travel time residuals, addressed the previous criticism brought up by Souriau et al. (1997). Through a similar joint inversion for IC rotation and heterogeneous structure, Song (2000a) confirmed a robust lateral velocity gradient in the IC and found the super-rotation to be between 0.3°/yr and 1.1°/yr. On the other side of the argument, Poupinet et al. (2000) demonstrate that the mislocation of earthquakes can significantly bias the estimates of IC rotation. Importantly, they did not deny the existence of IC rotation; but given the mislocation issue, they limited the differential rotation to no more than 1°/yr, if it exists at all.

5.3 Towards Complexity

Soon after the first seismological observations revealed a steady eastward rotation of the IC with respect to the mantle, it became possible to use the rotation rate to estimate IC viscosity. If the IC spins differentially in the presence of a strong gravitational torque imposed by the mantle, it has to adjust its shape as it spins. Buffett (1997) used this required balance to define two possible viscosity regimes. Firstly, if there is high viscosity ($v \geq 1.5 \times 10^{20}$ Pa s) and a slow rotation, the IC will lock itself to the mantle since the topography at the ICB would develop easier and join the mantle and core like interlocking teeth. As a result, the differential rotation would cease. Secondly, if the viscosity is low ($v \leq 3 \times 10^{16}$ Pa s), the 'victim' would be anisotropy, as low viscosity material does not favour the development of large crystals nor crystal alignment. At the beginning of the 2000s, after several studies presented in Section 5.2.3 challenged the ability to detect IC rotation, compromise was seen in a small rotation rate, possibly many times slower than originally reported. Anisotropy could still develop provided the viscosity of the IC is high, albeit at the cost of significant differential rotation. However, this did dampen interest in IC rotation amongst the seismological community.

5.3.1 Slow Rotation or No Rotation at All?

Not long after the first body wave travel time studies claimed to detect IC rotation, the normal mode community attempted the same. Normal modes can efficiently be used to probe for long period patterns since the corresponding splitting functions are radial averages relatively insensitive to short scale heterogeneity, unlike body waves. They are also sensitive to a rotation of a rigid body that takes place in the planet's interior. In principle, under ideal conditions where mantle structure is known and can be corrected for, and OC structure can be neglected, the rotation of splitting functions should correspond to the rotation of the IC (Figure 5.2b). Thus, if we can obtain normal mode splitting functions sensitive to IC at two different points in time and estimate the relative shift, we can deduce the rotation rate of the IC.

Sharrock and Woodhouse (1998) were the first to detect IC rotation using normal modes. Unlike Song and Richards (1996) and Su et al. (1996), they assumed that the fast axis of IC anisotropy is aligned with the rotation axis of the Earth. Because IC anisotropy has a zonal symmetry in alignment with the Earth's rotation, the zonal terms of splitting functions will be completely insensitive to IC differential

Table 5.1 *Summary of IC rotation studies with some model characteristics.*

Source	Method/data type	Character of rotation	Magnitude[a] of rotation
Song and Richards (1996)	anisotropy tilt PKP(BC)−PKIKP	constant eastward	$1.1°/yr$
Su et al. (1996)	anisotropy tilt PKIKP from ISC	constant eastward	$3°/yr$
Creager (1997)	velocity gradient PKP(BC)−PKIKP	constant eastward	$0.2-0.3°/yr$
Souriau et al. (1997)	sph. harm. expansion PKP(BC)−PKIKP	undetectable	
Souriau (1998b)		undetectable	
Souriau (1998c)	velocity gradient PKP(BC)−PKIKP	constant eastward	$\leq 1.1°/yr$
Sharrock and Woodhouse (1998)	normal mode splitting functions	constant westward	up to $-2.5°/yr$
Laske and Masters (1999)	normal modes splitting functions	constant eastw. or westw.	within ±0.2
Souriau and Poupinet (2000)	velocity gradient PKP(BC)−PKIKP	constant eastward	$\leq 1.0°/yr$
Poupinet et al. (2000)	earthquake doublets PKIKP	constant eastw. or westw.	within ±0.2
Song and Li (2000)	velocity gradient PKP(BC)−PKIKP	constant eastward	$0.6°/yr$
Vidale et al. (2000)	detecting variation in PKiKP scattering	constant eastward	$0.15°/yr$
Collier and Helffrich (2001)	velocity gradient PKP(BC)−PKIKP	constant or oscillatory	0.45 ± 0.25 to 0.74 ± 0.29 T=280 day
Laske and Masters (2003)	normal modes splitting functions	constant eastward	0.11 ± 0.13
Li and Richards (2003)	velocity gradient earthquake doublets	constant eastward	$0.4-1.0°/yr$
Zhang et al. (2005)	velocity gradient earthquake doublets	constant eastward	$0.3-0.5°/yr$
Vidale and Earle (2005)	detecting variation in PKP scattering	constant eastward	$0.05-0.1°/yr$
Wen (2006)	earthquake doublets PKiKP, PKIKP	eastw.+ICB topogr. or IC rapid growth	unspecified

Table 5.1 *(cont.)*

Source	Method/data type	Character of rotation	Magnitude[a] of rotation
Koper and Leyton (2006)	PKP precursors and coda interferometry	jerky rotation	unspecified
Zhang et al. (2008)	velocity gradient earthquake doublets	constant eastward	few $0.1°/yr$
Lindner et al. (2010)	diff. travel times PKP(BC)−PKIKP Bayesian inversion	constant eastward or accelerating	$0.39 \pm 0.22°/yr$ $0.24 - 0.56°/yr$ in last 55 yr
Waszek et al. (2011)	E-W hemis. shift PKiKP−PKIKP	constant eastward	$0.1 - 1°/10^6 yr$
Tkalčić et al. (2013b)	velocity gradient earthquake doublets	mean eastward + oscillatory	$0.25 - 0.48°/yr$
	Bayesian hierarchical	decadal	$T{=}1.0°/yr$

Note: [a]IC differential rotation magnitude is given relative to the rotation of the mantle and crust, which are considered to be rotating as a rigid body. The eastward rotation is positive (faster than the mantle) and the westward rotation is negative (slower than the mantle).

rotation. Although the splitting functions of IC sensitive normal modes are dominated by zonal terms, they are not completely void of non-zonal terms, which are sensitive to IC rotation. Therefore, Sharrock and Woodhouse (1998) removed the zonal terms from the splitting functions before comparison.

Sharrock and Woodhouse (1998) computed splitting functions for selected normal modes for two time intervals: 1978–1982 and 1993–1996, noting that the number of earthquakes recorded in 1978–1982 is smaller and that data are of poorer quality. By varying the relative weight of each dataset and comparing the resulting variances, they conclude that the two datasets for IC sensitive modes ($_3S_2$, $_9S_3$, $_{11}S_4$, $_{11}S_5$, and $_{13}S_3$) are mutually inconsistent, i.e. no optimal solution can be found that fits both datasets. The same analysis for normal modes sensitive exclusively to the mantle ($_0S_{13}$ and $_5S_6$) indicates, however, that the two datasets (from two chosen time intervals) are mutually consistent. A possible explanation for this contradiction is that IC structure varies in time. Indeed, when the zonal terms are removed from the two datasets, the most plausible solution is a westward rotation of the IC (Figure 5.6). Apart from $_3S_2$, this pattern is also obtained for individual modes, resulting in an average westward rotation of approximately 2°/yr. According to the authors, their study confirms the existence of time-dependence in the IC. However, it is possible that other time dependent factors contribute to the time dependence of IC rotation, as IC rotation alone cannot explain the inconsistencies between the two datasets taken from different time intervals.

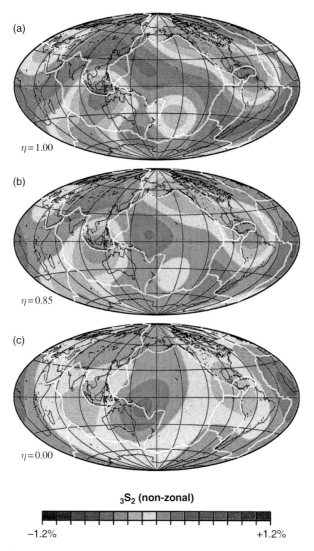

(a)

$\eta = 1.00$

(b)

$\eta = 0.85$

(c)

$\eta = 0.00$

$_3S_2$ (non-zonal)

-1.2% $+1.2\%$

Figure 5.6 Splitting functions for the mode $_3S_2$, derived from two groups of earth-quakes. The first group comes from the time interval 1977–1982, while the second group comes from 1993–1996. The parameter η defines the relative weight of data from each of the two time intervals analysed. When $\eta = 1.0$ only the 1977–1982 data group is inverted, when $\eta = 0.0$ only the 1993–1996 data group is inverted. All intermediate values quantify relative mix of contributions from each group. Only even degrees of the harmonic expansion of the splitting function are dis-played; the zonal terms, unaffected by core rotation, have been removed. The maximum scale is 1.2% of the mode frequency. a) $\eta = 1.0$ (less recent); (b) $\eta = 0.85$ (intermediate); (c) $\eta = 0.0$ (more recent). Modified from Sharrock and Woodhouse (1998).

The results of Sharrock and Woodhouse (1998) were surprising; on one hand, they confirmed the time-dependence structure in the IC, but on the other hand they contradicted all previously published results for IC rotation based on travel times. Laske and Masters (1999) used a technique that did not depend on the earthquake source, a.k.a. receiver stripping and linear autoregression (see Section 5.2.1). The modes of harmonic degree one are only sensitive to structure of degree two, which previous studies showed is almost entirely zonal in the IC (see Figure 2.20). Since zonal structure is axisymmetric, these modes will be insensitive to rotation around the pole. However, higher degree modes are not zonal, and can therefore be utilised to detect this rotation. After analysing the IC sensitive mode $_3S_2$ for the time intervals 1977–1985 and 1994–1998, Sharrock and Woodhouse (1998) determined that different components of splitting functions rotate with a different speeds and directions. The weaker, zonal degree two component rotates eastward, while the stronger, zonal degree four component rotates westward. This contradicts rotation of the IC as a rigid body. When they required that the IC rotate as a rigid body (i.e. the splitting functions of individual modes have to all rotate at the same rate), they obtained a mean rotation rate of almost zero ($0.01 \pm 0.21°$/yr). This rate contradicted all the previous estimates from body wave analysis except that of Creager (1992).

Laske and Masters (2003) later tried forcing all normal modes and all events in a given time window for a given mode to yield the same rotation rate, which is a plausible assumption if the IC rotates as a rigid body. The obtained IC rotation rate of the IC was relatively insensitive to the tomographic model used to correct for mantle structure, which was a positive development. However, even in the most optimistic scenario, the obtained differential rotation rate was too small to be reconciled with the results from body wave studies. Laske and Masters (2003) showed the differential rotation rate to be insignificant, at only $0.11 \pm 0.13°$/yr. This did nothing to unite the divided community, one half of which was convinced that these seismological methods were sophisticated enough to measure this captivating phenomenon taking place in the Earth's centre (e.g. Richards et al., 1998) and the other saw it as 'the last nail in the coffin' to the seismological attempts to observe it.

Apart from normal modes, scattering was also used to detect IC rotation. Vidale et al. (2000) used scattered waves from nuclear tests at Novaya Zemlya recorded on LASA array in Montana. They first detected IC scattering in the slant stacks from nuclear explosions. Then they simulated the distribution of change in arrivals between 1971 and 1974 by simulating a scattered wavefield, which affects the interference patterns and arrival times of the scattered energy. The best fit was obtained for a shift of $0.45°$, giving an IC rotation rate of $0.15°$/yr, in agreement with Laske and Masters (1999). At the same time, Isse and Nakanishi (2002) studied 27 years of data recorded by station SYO in Antarctica, and estimated a rotation of $0.2°$/yr or less, which was consistent with Vidale et al. (2000).

Vidale and Earle (2005) extended their coda technique, previously applied to PKiKP waves, to IC-grazing PKP waves from nuclear tests on Mururoa Island and recorded on the NORSAR array. Their results were consistent with a slow (approximately 0.05–0.1°/yr) eastward rotation of scatterers in the IC.

Xu and Song (2003) analysed the ray path from the SSI to recording stations in China. This path did not produce an acutely 'polar' geometry (ξ was 40°), however, the method depended on the position of the velocity gradient in the IC. This is a well-documented path since the Beijing network had been in operation from 1966, so detection hinged upon how well the velocity gradient in the IC could be modelled. Repeating the procedure of Creager (1997), they obtained a rotation rate of 0.4 ± 0.12°/yr in the eastward direction.

Sun et al. (2006) confirmed some earthquake mislocation issues with data from the SSI earthquakes recorded in Alaska, but concluded that IC super-rotation is still the best explanation for the anomalous data. This was accompanied by another study, an analysis of SSI event pairs of small spatial separation recorded at station COL in Alaska, which evinced variable IC rotation (Song and Poupinet, 2007). The technique requires both PKP(BC)−PKIKP and PKP(AB)−PKIKP differential travel times but not a precise knowledge of the earthquake locations.

Waszek et al. (2011) found that the E−W boundary shifts eastward with depth, in agreement with a slow IC super-rotation of $0.1–1$°$/10^6$yr. So deeper material, and therefore older material if the growth is from bottom to top, will be slightly displaced eastward from the original longitude at which it solidified. Waszek et al. (2011) attempted to reconcile the seismologically observed W−E dichotomy with the slow rotation. The larger rates of rotation observed seismologically were interpreted as temporary deviations from the average rate.

5.3.2 Earthquake Doublets

An earthquake doublet is a pair of earthquakes producing waveforms with a high level of similarity when recorded at the same station. This similarity is a sign that the waves were initiated at the same location and that they traversed the same Earth structure before being recorded at the same station. Additionally, this means that the earthquake source mechanisms, and therefore the energy radiation patterns, are very similar and that the separation between their foci is smaller than a typical wavelength at which the waveforms are observed.

Detecting Doublets

We can devise a search algorithm to compare all waveform records in a given database and identify doublets, triplets, or multiplets with sufficiently high cross-correlation coefficients to merit closer analysis of waveforms to confirm their status as repeating earthquakes. Repeating earthquakes are useful in many ways, for example to monitor changes in Earth structure on local and regional scales due to earthquakes or volcanic eruptions. High frequency waveforms in particular have significant sensitivity to small changes in Earth structure. However, using the same phenomenon on a global scale to monitor changes in IC structure takes a considerable scientific imagination and enthusiasm, if not scepticism at first. This is because at teleseismic distances many other factors come into play: the ray paths are long, and noise becomes an important factor, with attenuation and geometrical spreading often preventing the observation of PKP waves for small events on noisy stations. It seems that only an incredible stroke of luck would allow one to find a 'wiggle to wiggle' match between two teleseismic waveforms.

Broadband waveforms originating from large earthquakes contain information about Earth structure on many scales. If short scales are eliminated and we focus on major, long-scale features using long periods, we can perhaps reduce the undesired effects of noise and attenuation. We can even speed up our search algorithm to identify repeating earthquakes based on their long periods, which requires less computer memory. As a test, let us consider earthquakes from the South Sandwich Islands region (SSI). The chosen recording station must have continuous recordings (so that it can capture as many events as possible over a given time interval) and a close proximity to SSI (so that the surface waves of even moderate events are detectable). The closer the station is to the source, the shallower the ray paths will sample, eventually avoiding the Earth's core entirely. There are not many suitable stations for the SSI earthquakes due to their remote location. There are some permanent stations in Antarctica, but seasonal variations in temperature, ice structure, and ice thickness make those stations too noisy and potentially affect the paths of surface waves, so an initial search for similar waveforms recorded at these stations was unsuccessful. Fortunately, there are also appropriately positioned stations in South America, and one of them, the station NNA in Peru, is located in a quiet environment ideal for our search. Figure 5.7 shows a doublet from the SSI recorded on station NNA. For a station located about 60° (more than 6000 km) away from the source , it is incredible to observe such a perfect match between two waveforms, as can only be expected from true doublets.

The doublets from the SSI were first reported and used by Poupinet et al. (2000) to scrutinise the seismological detection of the super-rotation of the Earth's IC. In that study no significant temporal changes were observed in the travel times of the doublets from the SSI to Alaska path within the resolution of the method, which the authors set at 0.05 s. They ascribed the previously observed variations in travel times (Song and Richards, 1996) to hypocentre mislocations. This denial of the detectability of IC rotation took an unexpected turn, however, and opened onto a new avenue of rebutting previous criticisms that the travel times of PKIKP waves are significantly affected by mislocation errors and contamination from small-scale

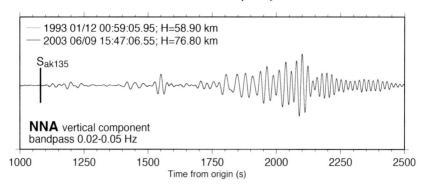

Figure 5.7 Earthquake doublet 01/12/93–06/09/03 (doublet 14 in Figure 5.13) from the SSI region recorded at Nana station, Peru, bandpass filtered to focus on long periods, unlike the short period focus of Figure 5.8. The earthquake parameters are shown in the upper left corner. The seismogram for the earlier earthquake is in grey and the seismogram for the more recent earthquake is in black. Traces are clipped before S-wave arrivals, aligned on the S-wave phase, filtered between 0.02 and 0.05 Hz to minimise effects of short scale structure, and overlapped. The similarity of the two earthquakes is flaunted by the fact that the two traces cannot be distinguished for the entire S wave coda and Rayleigh waves at these periods, except near the S wave onsets. See Figure 5.10 for compressional wave records and measurements for the same doublet, recorded on station College, Alaska.

inhomogeneities. The earthquake doublets occurring in the SSI region observed on station COL in Alaska presented a unique opportunity to detect temporal changes associated with the differential rotation of the IC. Song (2000b) used earthquake doublets to rule out bias in IC rotation detection due to earthquake mislocations. The study showed that the choice of reference model affects travel time residuals, and they alleged that the subjective choice of doublets led to the inability to detect IC rotation.

Using earthquake doublets, a significant and steady differential eastward rotation of the IC was confirmed to be approximately 0.4–1.0°/yr by Li and Richards (2003) and 0.27–0.53°/yr by Zhang et al. (2005). We show two examples of earthquake doublets from the SSI region recorded by station College in Alaska (COL) at the epicentral distances corresponding to the arrivals of PKP waves. The first doublet (Figure 5.8) is the doublet previously featured in Figure 5.7; however, it is now bandpass filtered between much higher frequencies (0.6–2.0 Hz). This doublet was reported as the Doublet 14 in Tkalčić et al. (2013b). The time separation between the two events in this doublet is just under ten years. Again, a high level of similarity is observed; the waveforms match each other in a 'wiggle by wiggle' fashion. The second doublet (Figure 5.9) spans a little over seven years and was reported as Doublet 11 in Tkalčić et al. (2013b). Interestingly, from the time of the first paper by Li and Richards (2003) to that time of the second paper by Zhang et al. (2005),

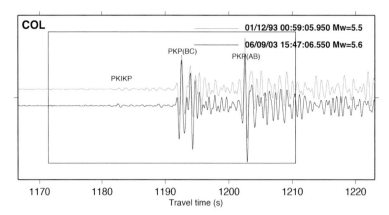

Figure 5.8 Earthquake doublet 01/12/93 – 06/09/03 (doublet 14 in Figure 5.13) from the SSI region recorded on station COL (College), Alaska, bandpass filtered between 0.6 and 2.0 Hz to highlight dominant frequencies of PKP waves around 1 Hz. The earthquake parameters are shown in the upper right corner. The seismogram for the earlier earthquake is in grey and the seismogram for more the recent earthquake is in black. Traces are clipped before the PKP-wave arrivals, aligned on the PKP(BC)-phase, and shown one above another. The window used for differential travel time measurements is marked by a rectangle and enlarged in Figure 5.10. The long period traces of the same doublet recorded on station Nana, Peru, are shown in Figure 5.7

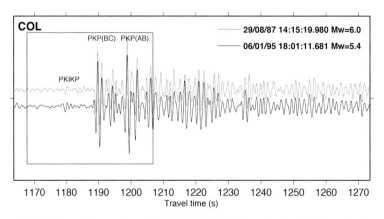

Figure 5.9 Earthquake doublet 29/08/87–06/01/95 (doublet 11 in Figure 5.13) from the SSI region recorded at College station, Alaska, bandpass filtered between 0.6 and 1.2 Hz to highlight the dominant frequencies of PKP waves. See captions of Figure 5.8 for explanation. The window used in the differential travel time measurements is marked by a rectangle and enlarged in Figure 5.11. Adopted from Tkalčić et al. (2013b).

the number of identified doublets from the SSI region recorded on station COL increased from 1 to 17. Although these observations supported a steady differential super-rotation of the IC, they did not ease the continued controversy between body waves and normal modes results.

Analysing Doublets

Once the doublets, triplets, and multiplets are identified, the more difficult process of measuring the differential travel times remains. The cross-correlation measurements here will be equivalent to the differential travel time relationship of Equation 1, but we will use observed instead of theoretically predicted travel times. There are at least three different ways to quantify the arrival time differences of the phases. Our ultimate goal is to measure the time difference between the phase PKP(BC) (insensitive to IC structure) and the phase PKIKP (sensitive to IC structure) for both the earlier and the later event of each doublet and then compare these two differences. This can be expressed as

$$\Delta_{\delta t} = [PKP(BC) - PKIKP]_2 - [PKP(BC) - PKIKP]_1, \quad (5.6)$$

where subscript 1 refers to the waveform of the earlier event and subscript 2 refers to the waveform of the later event.

If δt is invariant in time ($\frac{\partial}{\partial t}\delta t = 0$) and the two phases travel through the same medium, then there is no indication of a change in IC structure. If, however, δt is significantly different from zero and beyond the resolution limit of the method (recall that Poupinet et al., 2000, set the resolution limit equal to approximately 0.05 s), we can claim to have detected differential rotation and calculate its rate as shown in Section 5.2.1.

The first method of calculating $\Delta_{\delta t}$ (Method 1) is to align the PKP(BC) phases of the two waveforms. Assuming $PKP(BC)_1 = PKP(BC)_2$, Equation 5.6 gives

$$\Delta_{\delta t} = PKIKP_1 - PKIKP_2. \quad (5.7)$$

The advantage of Method 1 is that it represents a direct comparison of PKIKP waveforms without the need for cross-correlating the PKP(BC) and PKIKP waveforms, which can be problematic if PKIKP is significantly attenuated by IC structure and not appropriately corrected. The second method (Method 2) follows Equation 5.6. A positive $\Delta_{\delta t}$ means that the PKIKP wave of the later event arrived earlier, indicating a super-rotation of the IC if the velocity gradient is positive i.e. PKIKP waves sample progressively faster medium with time. Lastly, a variant of Method 1 consists of aligning the doublet traces according to the PKP(BC) phases, picking the first visible maximum of the PKIKP arrivals, and computing the time between the two picks. We refer to it as Method 3, but due to larger uncertainties this method is rarely used, especially if full waveform comparison is possible.

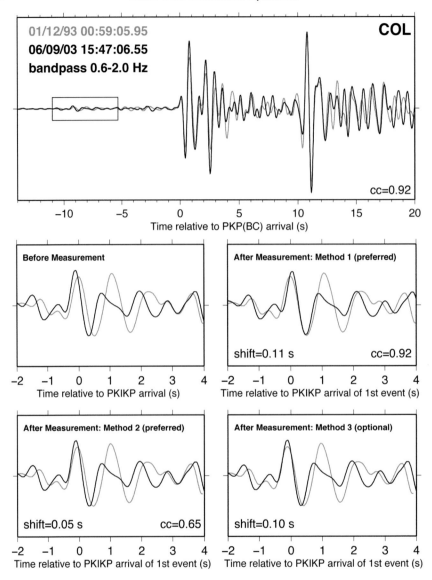

Figure 5.10 Differential travel time measurements for earthquake doublet 01/12/93–06/09/03 (doublet 14 in Figure 5.13 and shown in Figures 5.8 and 5.7) from the SSI region recorded at College station, Alaska, bandpass filtered between 0.6 and 2.0 Hz to highlight the dominant frequencies of PKP waves. (Top) Overlapped PKP waveforms of the earlier (grey) and more recent event (black), containing all three branches of PKP waves aligned on PKP(BC) phase. The time interval around the PKIKP waves is marked by a rectangle. Also shown is the bandpass filter applied prior to measuring (for example, bp c 0.8 1.5 means: bandpass filter with corner frequencies of 0.8 and 1.5 Hz) and the obtained cross-correlation coefficient for the entire waveform. (Middle) The waveforms around the PKIKP onsets are incised by a rectangle in the top panel and are enlarged to clearly show the alignment of the PKIKP phases before (left) and after (right)

We show two examples of the earthquake doublets analysis in Figures 5.10 and 5.11. All three methods are illustrated in each figure, and the resulting $\Delta_{\delta t}$ (referred to as 'shift') is given in each figure. Subtle differences among the methods are apparent. Particularly significant is the difference between Method 3 and Methods 1 and 2 for the doublet $29/08/87 - 06/01/95$, which attests to the larger uncertainty of Method 3.

The use of earthquake doublets to study IC rotation continued throughout the 2000s. Wen (2006) published observations of PKiKP waves in Asia (Russia and Kazakhstan) using the SSI doublets. The study argues for either a differential rotation of the IC with pronounced topography or a growth of the ICB, where topography changes from 0.98 to 1.75 km over ten years. A slope in ICB topography would have a similar effect on the travel times of PKiKP waves. Zhang et al. (2008) reported new doublets for SSI, Tonga–Fiji, and Kuril islands earthquakes, including some triplets and multiplets. Some of these exhibited temporal changes, others did not. They attributed this discrepancy to the differences in source-receiver geometries.

5.4 Complicated Rotational Dynamics of the IC

5.4.1 Non-Steady Rotation

Detection of IC rotation faced challenges since the pioneering studies of the mid 1990s. Much of the scepticism related to the much smaller magnitude of time measurements required for IC rotation analysis in comparison with other studies, for example on anisotropy. Typically on the order of a tenth of a second or less, these time differences are arguably below the resolution limit of the methods we employ. However, the ability to detect IC differential rotation has gained support since the emergence of the earthquake doublets method, not because of the improvements the method introduces in the measurements themselves, but because of the reduction of uncertainties related to our assumptions.

Indeed, it is not the measurement quality that needed improvement, but rather the assumptions used to infer the rotation rate that could improve. For example, the most serious uncertainty, earthquake mislocation, was elegantly circumvented

---

Caption for Figure 5.10 (cont.) the time shift obtained by cross-correlation (Method 1). Also shown are the cross-correlation coefficient and the shift in seconds; (Bottom) Alignment of PKIKP phases after the shifts obtained from Methods 2 (left) and 3 (right). Also shown are the cross-correlation coefficient and the shift in seconds for Method 2 and the shift in seconds for Method 3 (Tkalčić et al., 2013b). For the explanation of different methods of measurement, see the text.

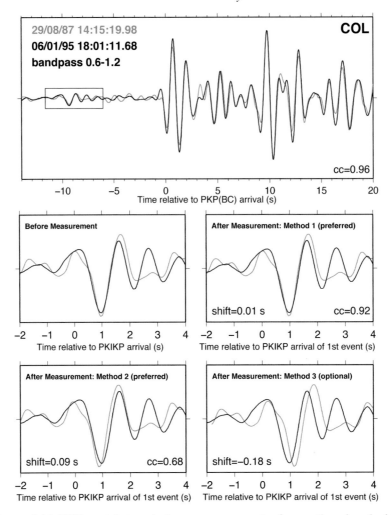

Figure 5.11 Differential travel time measurements for earthquake doublet 29/08/87–06/01/95 (doublet 11 in Figure 5.13 and shown in Figure 5.8) from the SSI region recorded at College station, Alaska, bandpass filtered between 0.6 and 1.2 Hz to highlight the dominant frequencies of PKP waves. See captions of Figure 5.10 for the explanation, and for more details on different methods of measurements, see the text. Modified from Tkalčić et al. (2013b).

through the use of earthquake doublets. Thus, there is no doubt that the new method paved a new path in our journey towards understanding IC rotation. Another improvement awaited, which was related to the process of geophysical inference and the assumption of how the IC behaves. The majority of the studies assumed a steady rotation of the IC. The addition of newly discovered doublets raised doubts about the existence of a steady rotation of the IC, thus prompting a loosening of the requirement for constant rotation.

To explain the doubts surrounding a steady IC rotation, let us consider an example where two doublets of a similar time separation are plotted against absolute time. Instead of representing $\Delta_{\delta t}$ (Equation 5.6) as a point in time, we represent a doublet using two points in time connected by a line. $\Delta_{\delta t}$ is now indicated by the separation of the two end points on the vertical axis, though its position along this axis is irrelevant. A positive slope means that PKIKP waves traversed the IC faster in more recent years. This is shown in Figure 5.12 for all identified doublets

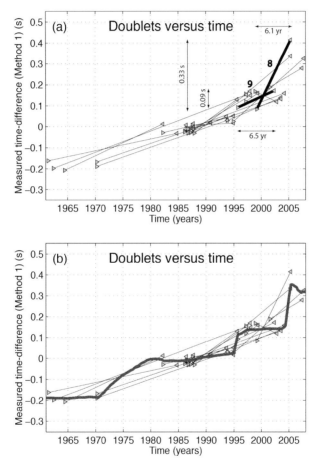

Figure 5.12 (a) Earthquake doublets from the SSI region recorded on station COL. The time-difference $\Delta_{\delta t}$ (Equation 5.6) measured for each doublet using Method 1 (see the text for explanation of the method) is plotted as a function of absolute time. Triangles mark the time of each earthquake of a doublet on the horizontal axis. Lines are used to join pairs of events making a doublet. The two highlighted doublets (thick lines) are also shown in Figure 5.13 and featured in Tkalčić et al. (2013b) as doublets 8 and 9. Also shown are the measurements (in seconds) and time lapse (in years) for these two doublets.

from the SSI as recorded at station COL, with a particular emphasis on two doublets (identified as doublets 8 and 9 in Figure 5.13a). From Figure 5.12a it is clear that two doublets of the same time separation can exhibit different slopes. Thus, what if instead of inverting for a constant slope over the entire time interval (linear fit; steady rotation), we could relax our requirement for a constant slope and allow for variable rotation? What if we could treat the change of slope as a function of time as an unknown? How would we need to parameterise our model? How many additional parameters would be required? How would we justify such a complex parameterisations as opposed to a simple linear fit? We will see in this chapter how these questions were tackled.

5.4.2 Observations of Non-Steady IC Rotation

It is worth recalling here that Su et al. (1996) published a pioneering work on IC rotation, claiming that they detected not just a steady rotation, but also a significant anomaly in the time interval 1969–1973, which they related to the geomagnetic jerk of 1969–1970. Steady rotation dominated subsequent seismological predictions, probably because the idea of an oscillatory rotation was suppressed by a polemic on a more basic level, about whether or not IC differential rotation can even be detected.

Finally, Collier and Helffrich (2001) presented weak evidence, as they called it, for rotational oscillation on a timescale of 280 days. They identified and characterised a velocity gradient in the IC and used it to estimate the rotation rate according to the method described in Section 5.2.1 and shown in Figure 5.2b. They considered both a straight line and a sinusoidal function when trend-fitting their data. They obtained a variance reduction of about 50 per cent and ruled out a rotation rate greater than $1°/yr$ at the 95 per cent confidence level. They also reported a high IC viscosity ($v = 3.9 \times 10^{19} Pas$) based on the rotation rate. Such high viscosity would prohibit deformation of the IC and would require the IC to be gravitationally locked to the mantle.

Koper and Leyton (2006) examined the precursors and coda of PKP waves from a well-known doublet featured in Figures 5.7, 5.8, and 5.10. This doublet spans the time interval 1993–2003 and was introduced by Zhang et al. (2005) and used later by Tkalčić et al. (2013b) as Doublet 14. Koper and Leyton (2006) also identified a doublet not used previously that spanned the time interval 1993–2001. An interferometry method (Snieder, 2004) was applied to these doublets to monitor IC motion. A high correlation between the traces in the windows before (precursory window), during (PKP window), and after the main PKP arrivals (coda window) would indicate a static IC. High correlation only within the precursory window indicates that the IC has moved. They found strong evidence for rotation using the

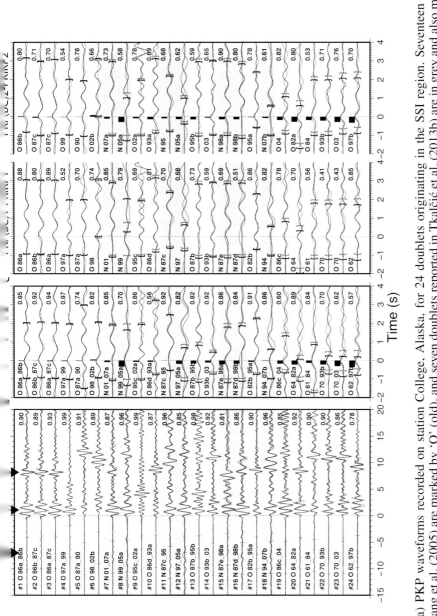

Figure 5.13 (a) PKP waveforms recorded on station College, Alaska, for 24 doublets originating in the SSI region. Seventeen doublets reported in Zhang et al. (2005) are marked by 'O' (old), and seven doublets reported in Tkalčić et al. (2013b) are in grey and also marked by 'N' (new). The traces (grey for the earlier events and black for the later events) are aligned at the PKP(BC) phase. They are sorted by time separation increasing from top to bottom. (b) Method 1: Enlarged PKIKP segments for each doublet marked by ticks in (a). The segments indicated by brackets have been aligned using cross-correlation. The traces have been shifted so that the onset of the PKIKP arrival of the earlier event of each doublet is roughly aligned. The black horizontal bars indicate the measured difference of PKP(BC)−PKIKP travel times. (c) Method 2: Enlarged PKIKP and PKP(BC) segments of the first event of a doublet (left) and the second event of a doublet (right). The traces indicated by brackets have been aligned using cross-correlation. The black horizontal bars indicate the measured difference of PKP(BC)−PKIKP travel times. Individual measurements for doublets 11 and 14 are shown in Figures 5.10 and 5.11. Modified from Tkalčić et al. (2013b).

first doublet and no evidence for rotation using the second doublet. This simultaneous evidence of changing structure in the IC over a large time window and non-changing structure over a shorter time window led them to speculate that the IC might display a jerky rotation rather than a smooth, constant super-rotation.

Lindner et al. (2010) separated the time-independent effects of mantle structure from the time-dependent effects of IC structure. They performed a Bayesian inversion without using earthquake doublets. They concluded that the rotation rate was $0.39 \pm 0.22°$/yr, and found an acceleration from $0.24°$/yr to $0.56°$/yr in the 15 years preceding their work.

A commonality between the studies of Su et al. (1996) and Collier and Helffrich (2001) was that they used a large dataset of differential travel times, but not earthquake doublets. Nor did the later work of Lindner et al. (2010) use them, though doublets had been studied extensively by then. What temporarily hampered the use of earthquake doublets was the notion that all doublets along the most significant ray paths for which there is a continuous recording had already been discovered. Moreover, Mäkinen and Deuss (2011) argued that earthquake doublets often have lower signal-to-noise ratios than individual large events, making measurements difficult. The same authors concluded that the doublets used by Zhang et al. (2008) as a further evidence of IC differential rotation cannot resolve super-rotation of the IC. This assertion was based on the observation that the measured temporal variation of the PKP travel time residuals was different for different IC sampling geometries, whereas they should all be the same if the IC rotates as a rigid body. However, Tkalčić et al. (2013b) performed a systematic analysis of the SSI earthquakes recorded on station College, Alaska and increased the existing library of doublets along this path by 40 per cent (Figure 5.13).

5.4.3 A Shuffling Rotation of the IC

We saw in Section 5.4.1 that if doublets are considered according to absolute time, the time difference $\Delta_{\Delta t}$ requires a best-fit curve of variable slope (Figure 5.12a). Some of the observations presented in Section 5.4.2 undoubtedly confirm that the IC rotation is non-steady. With each newly discovered doublet, the dataset increases, and the fit to the data is better constrained. Is it now possible to use a more complex parameterisation to recover a full time-history of the slope variations? Is it possible to prove this solution is preferred over a linear fit? Luckily, the answer to both of these questions is affirmative due to recent advances in geophysical inverse theory within a Bayesian framework.

Within a Bayesian framework, the solution sought is not a single best-fit model, but a large ensemble of models that are distributed according to a posterior probability density function. If $\mathbf{m}$ is the vector of the model parameters and $\mathbf{d}$ is the vector defined by the set of measured data, then the posterior probability density function can be expressed as $p(\mathbf{m}|\mathbf{d})$, i.e. the probability of the model parameters given the measurements. Bayes' theorem (Bayes, 1763) combines prior information about the model, $p(\mathbf{m})$, with a likelihood function, $p(\mathbf{d}|\mathbf{m})$, or literally the probability of the observed data given the model, to give the posterior probability density function

$$p(\mathbf{m}|\mathbf{d}) \propto p(\mathbf{d}|\mathbf{m}) \times p(\mathbf{m}), \tag{5.8}$$

where $a|b$ means a given, or conditional on, b, i.e. the probability of having a when b is fixed. A Bayesian transdimensional inversion framework (see Appendix D) allows more objective choices when finding a causal relationship between the measured time-differences, $\Delta_{\delta t}$, for each doublet and the time lapse (about 55 years in total). Numerous potential solutions are generated using the reversible jump Monte Carlo algorithm (see e.g. Green, 1995). The parameterisation, and therefore the number of unknowns, are left as variables, meaning that rather than imposing a linear fit to $\Delta_{\delta t}$ as a function of time, we can invert for the expected slope, its change in time, and the occurrence of likely change points. This approach can be further extended using a hierarchical model (see Appendix D), which also includes the degree of data uncertainty as an unknown (e.g. Malinverno and Briggs, 2004).

We can obtain a fit to the measured time-differences, $\Delta_{\delta t}$, as a function of time with the accompanying probability of an abrupt change at each point along this function. Figure 5.12b shows a solution averaged over 300,000 model realisations. The variation of the time differences curve is complex, with the most probable increases in slope occurring during 1971, 1994, and 2004. The uncertainty on each doublet datum is assumed to be proportional to the inverse of the cross-correlation coefficient associated with the doublet, with the constant of proportionality serving as the unknown in the problem. Assuming known heterogeneity in the IC, $\frac{\partial v}{\partial L}$ (%/°), along the sampling IC profile, L, the measurements of $\Delta_{\delta t}$ expressed as a function of absolute time can be used to estimate the differential IC rotation rate. The slope can be calculated at each point in time by differentiating the curve shown in Figure 5.12b. This is shown in Figure 5.14a. The slope can then be converted to an IC rotation rate, α, using Equation 5.5.

Figure 5.14b shows the recent history of the IC rotation rate based on the known marker, $\frac{\partial v}{\partial L}$, obtained in a previous study (Song, 2000a). The velocity gradient, $\frac{\partial v}{\partial L}$, depends on how crustal and mantle structure is assumed to effect the measured time differences, $\Delta_{\delta t}$. With regards to the IC rotation history, instead of a steady rotation, the IC tends to accelerate to rate of several degrees per year over the period of one to two years and decelerates back to a negligible rotation over the course of between five and 15 years. The prominent acceleration during 2003 followed by a prominent deceleration during 2004 is surprising, and may be considered an artefact of the

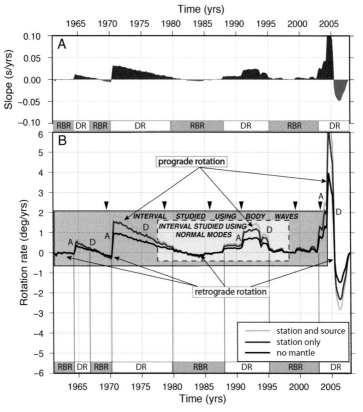

Figure 5.14 (a) The slope parameter (for explanation, see Equation 5.5) as determined by Bayesian inversion, in which the total time interval is divided into a variable number of constant (one-dimensional) Voronoi partitions separated by sharp discontinuities. The unknown parameters in the inversion are the number of Voronoi nuclei, their position along the time axis, and the regression value within each cell. Blue areas indicate eastward rotation whereas red areas indicate westward rotation of the IC with respect to the mantle. (b) Differential rotation rate of the IC (Equation 5.5) as derived from the slope for three different values of the velocity gradient (depending on the mantle model used for corrections) defined in Song (2000a). The dark grey and light grey rectangles delineate the time intervals used in the study of doublets (Zhang et al., 2005) and normal modes (Laske and Masters, 2003). The areas under these curves were integrated to estimate the total shift. The position of the reported geomagnetic jerks is indicated by black inverted triangles. 'DR' stands for 'differential rotation' and 'RBR' stands for 'rigid body rotation'. (A black and white version of this figure will appear in some formats. For the colour version, please refer to the plate section.)

Voronoi parameterisation. Despite the fact that the Voronoi cell parameterisation can recover a smooth function (for example, see synthetic experiments in Tkalčić et al. (2015)), the changes in IC rotation shown in Figure 5.14 come across as overly abrupt.

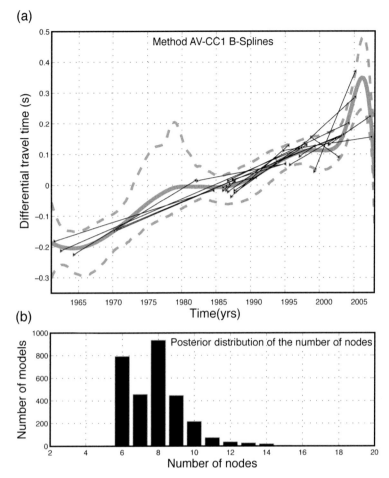

Figure 5.15 (a) The data associated with each doublet (black triangles are measured values for each earthquake of a doublet, joined using straight lines) are fitted using a variable slope model via a Bayesian transdimensional inversion and is the average of an ensemble of sampled models. The grey thick line is the expected solution model (normalised to zero mean). The dashed grey line is an average solution ±1 standard deviation. (b) Posterior distribution for the number of B-spline nodes across the ensemble solution. Modified from Tkalčić et al. (2013b).

Tkalčić et al. (2013b) used cubic splines instead of Voronoi cells and obtained similar results. Figure 5.15 shows the variable slope model resulting from averaging over the ensemble of collected models. It also shows the posterior distribution for the number of B-spline nodes across the ensemble solution. After converting from slope to rotation rate, the solution is similar to that obtained with the Voronoi parameterisation, although the corresponding slope and rotation rate curves are smoother (Figure 5.16). The average differential rotation rate of

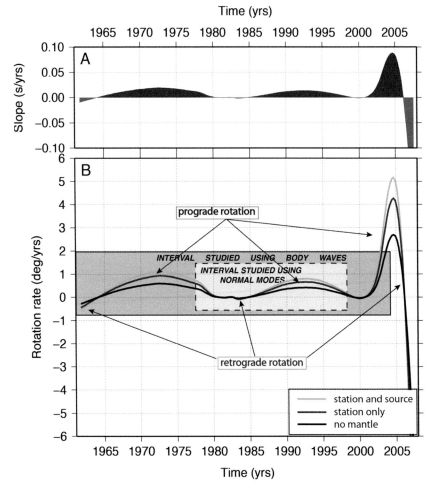

Figure 5.16 (a) The slope parameter (for explanation, see Equation 5.5) as determined by the Bayesian inversion of Tkalčić et al. (2013b), where the time interval is divided into a variable number of B-splines, which itself is a free parameter in the inversion. Blue areas indicate eastward rotation whereas red areas indicate westward rotation of the IC with respect to the mantle. (b) Differential rotation rate of the IC (Equation 5.5) as derived from the slope for three different values for the velocity gradient (depending on the mantle model used for corrections) defined in Song (2000a). The dark grey and the light grey rectangles delineate the time intervals used in the study of doublets (Zhang et al., 2005) and normal modes (Laske and Masters, 2003). The areas under these curves were integrated to estimate the total shift. Adopted from Tkalčić et al. (2013b). (A black and white version of this figure will appear in some formats. For the colour version, please refer to the plate section.)

0.25–0.48°/yr confirms an eastward rotation; however, decadal fluctuations of the order of 1.0°/yr are superimposed on that steady component. In practice, this means that the IC rotation rates vary in time, and that apart from eastward rotation, there were time intervals when the IC rotated westward with respect to the rest of the mantle, hence the term 'shuffling rotation' in the title of the study.

There are several immediate implications of this finding. Firstly, the fluctuation pattern of IC differential rotation reconciles discrepancies between normal mode and travel time observations, since the time intervals studied by both observational tools are characterised by different rotation rates of the IC. These time intervals are indicated by differently shaded rectangles in Figures 5.14 and 5.16. When the areas beneath the curves are integrated and divided by the time lapsed, the resulting average rates of differential rotation are in excellent agreement with the published rates from both travel time and normal mode studies. Secondly, the amplitude of fluctuation presents a new constraint on the gravitational coupling between the IC and the mantle (e.g. Dumberry, 2007; Dumberry and Mound, 2010). For example, in a recent study, Livermore et al. (2013) showed that the observed quasi-oscillatory nature of the IC may be driven by decadal changes in the magnetic field.

The LOD time series data feature rapid changes, some of which are well corre-lated with geomagnetic jerks (sudden change in the second derivative of the Earth's geomagnetic field), indicating that the Earth's core may be responsible for both phenomena (e.g. Holme and de Viron, 2005). In addition, the LOD data predict oscillation in the angular alignment of the IC (Buffett and Creager, 1999), thus supporting its differential rotation with respect to the mantle. Apart from four globally observed geomagnetic jerks in 1969, 1978, 1991, and 1999, there are two geomagnetic jerks, in 1986 and 2003, observed in only certain places in the world. Geomagnetic jerks are strongly dependent on the local field at the top of the core, and the fact that they are not always observed globally does not make them less significant (Bloxham et al., 2002). Interestingly, when Voronoi cells are used to parameterise the history of IC rotation (Tkalčić et al., 2013b) all three of the most dramatic changes in the rotation rate correlate with known geomagnetic jerks occurring around 1969, 1991, and 2003 (Figure 5.14). The other three com-monly reported geomagnetic jerks (1978, 1986, and 1999), however, do not show an obvious correlation with significant changes in IC rotation rate.

The correlation between the observed acceleration in the IC rotation rate and the observed geomagnetic jerks is statistically significant, and therefore unlikely to be coincidental. Although the physical origin of geomagnetic jerks and their connec-tion with the excitation of torques on the IC is not well understood (Bloxham et al., 2002), the observed correlation between the occurrence of geomagnetic jerks and the change in the rotation rate of the IC is likely a consequence of the rotational dynamics of the IC rather than other mechanisms such as changes of IC topography (Wen, 2006; Cao et al., 2007) or rapidly changing structure in the IC or OC (Mäkinen and Deuss, 2011). Changes in the fluid flow in the OC can entrain an axial rotation of the IC through electromagnetic stresses.

The characterisation of heterogeneous structure in the IC, and in particular in portions displaying velocity variations suitable as markers to measure the IC differential rotation rate, is of a high importance. Assumptions made about IC structure are crucial, because they so strongly influence results based on the marker method. IC velocities sampled by the SSI region to Alaska ray paths increase by approximately 0.5 per cent across a lateral extent of about 27°, which corresponds to about 450 km in the IC. Although this sharp gradient is well constrained, it is questionable whether the change in velocity over the entire length of the sampling profile, L, is constant. Tkalčić et al. (2013a) investigated the null hypothesis that the velocity gradient along the SSI to Alaska path in the IC is constant (i.e. δt is a linear function of azimuth) using transdimensional Bayesian inference, and found that in an ensemble of likely models of variable node points, the most likely ones contained only two nodes, which corresponds to a linear trend. The average model is shown as a red line in Figure 5.4; it contains a slight 'bend' near the azimuth of 312° due to contributions from less probable, higher order models. These results justify the assumption of a constant velocity gradient. It is important that future studies employ array analyses to further constrain the gradients in elastic properties of the IC, as this information is key to producing accurate geodynamical models.

Where do we go from here? A fluctuating rotation model is the only plausible model at the time that explains both body wave and normal mode observations. A potential way to further test the model would be to investigate the patterns of splitting functions over a continuous time interval, using overlapping time windows or similar approach. Until further validating analyses are performed, we must question how realistic the strong acceleration rate observed in recent years is. The answer might elude us until more doublets are discovered or another method invented.

6

The Limitations, the Obstacles, and the Way Forward

"If you can find a path with no obstacles, it probably doesn't lead any-where."

Anonymous

6.1 Limited Distribution of Seismic Sources and Receivers

Exploration seismologists estimate properties of the Earth's subsurface using reflecting or refracting parts of the wavefield generated by controlled sources such as dynamite, air guns, or seismic vibrators. They have control over the quantity of sources and receivers they deploy when investigating potential reservoirs of geothermal energy, coals, minerals, or hydrocarbon. In medical imaging, where the main purpose is to diagnose or treat a disease, there is a similar 'luxury', since the number of sources and receivers is controlled through sophisticated design. Even in helioseismology a favourable distribution of receivers can be achieved via proper experimental design, i.e. by orbiting around the Sun along a variety of paths. However, in terrestrial seismology there are fundamental reasons preventing complete volumetric sampling of the deep Earth by seismic body waves. Firstly, the majority of earthquakes large enough to be observed on the other side of the globe and utilised in IC studies occur in subduction zones within moderate lati-tudes. Secondly, the global distribution of seismic instruments is uneven. This is understandable given the uneven distribution of accessible landmass on the Earth's surface and the non-uniform distribution of wealth in the world. A large number of ocean bottom seismometers were installed recently; however, the recording of core-sensitive data is hampered by large amounts of noise in the 1 Hz range. The uneven distribution of events and instruments greatly limits the number of polar paths through the IC.

To illustrate how critical the lack of complete global sampling is for studies of seismic anisotropy in the IC, the angle between PKIKP waves and the rotation axis of the Earth ξ is computed for all hypothetical source-receiver pairs on the Earth's surface. Figure 6.1a helps illustrate that for a hypothetical earthquake located at the equator, the range of angles ξ that can be achieved by PKIKP waves traversing the IC to the recording stations at any given location on the Earth's surface is 70–90°. In order to achieve IC sampling by angles smaller than 30°, either the source or the receiver has to be located at latitudes beyond 60°. Figure 6.1b shows that to achieve the same kind of sampling in the centremost part of the IC (where PKP waves are almost antipodal), both the source and the receiver must lie at latitudes beyond 60°.

There are occasional seismic events at latitudes close to the Earth's poles, such as those in the South Sandwich Islands subduction zone. Other locations include the Macquarie Ridge, the Svalbard Sea, and Siberia. There were also nuclear explosions at high latitudes (such as those in Novaya Zemlya). Events with reverse and normal focal mechanisms radiate significant seismic energy vertically downward and thus produce seismic phases sensitive to the Earth's core, such as PKIKP waves. Strike-slip events tend to radiate small amounts of energy in the vertical, downward direction, thus producing weak PKIKP waves.

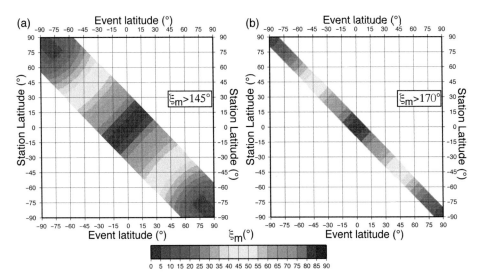

Figure 6.1 (a) Map of the smallest possible angle between PKIKP waves in the IC and the rotation axis of the Earth (ξ) for any given source-receiver pair at the Earth's surface under the restriction that $\Delta \geq 145°$, where Δ is epicentral distance. Source and receiver latitudes are plotted on the horizontal and vertical axes. Colours correspond to different values of ξ. White areas are source-receiver pairs for which the geometry of the PKIKP waves does not satisfy the above condition. (b) Same as (a) but under the restriction that $\Delta \geq 170°$. After Tkalčić (2015). (A black and white version of this figure will appear in some formats. For the colour version, please refer to the plate section.)

During the last few decades, the number of seismic instruments has increased dramatically. In the year 2000 there were about 150 stations within the Global Seismic Network. By the year 2012 there were several hundred, and for certain events and time intervals, more than 1000 stations with waveforms readily available online for download and analysis. The 1990s were the most prolific, and resulted in quite a few new stations capable of recording PKP phases, particularly from earthquakes originating in South America and the South Sandwich Islands region. However, this significant global expansion did not yield an equivalent increase in the number of global stations recording PKP phases, as most of the stations were not sufficiently antipodal relative to major seismogenic zones as to satisfy the condition $\Delta \geq 145°$ (for a detailed review, see Tkalčić, 2015).

6.1.1 PKIKP Waves

There are far fewer quasi-polar paths than quasi-equatorial paths (for the definition see Chapter 4), though the transition between polar and equatorial sampling is somewhat arbitrary. If we consider $\xi \leq 35°$ to represent quasi-polar paths, then Figures 6.2 and 6.3 clearly illustrate the uneven distribution of raypath geometries through the IC. These figures show an abundance of quasi-equatorial paths and a sparsity of quasi-polar sampling in both hemispheres of the IC. The South Sandwich Islands earthquakes produce anomalously advanced PKIKP travel times in the qWH of the IC (the bundle of dark blue ray paths clearly visible in Figures 6.2c and 6.3a). This anomaly has been used as an argument for a hemispherical nature of IC anisotropy in its upper portion (Chapter 4). Most of the IC, however, remains poorly sampled or not sampled at all.

The best global sampling of the IC is achieved in the qEH under the Indian Ocean (Figure 6.4a) and southeast Asia (Figure 6.4b) due to the close vicinity of major seismogenic zones in Indonesia, Tonga–Fiji, and Japan (earthquakes in Figure 6.4 are shown by red balls) and the fact that the PKIKP ray paths point steeply down from the source. There is, however, a lack of seismic stations providing continuous waveform data on the African and South American continents and Antarctica (stations in Figure 6.4 are shown by green balls). Large events in the Pacific and Atlantic oceans with a favourable radiation pattern (non strike-slip) are rare, so there is minimal sampling of the lowermost mantle and the IC in the Pacific Ocean (c) and the Atlantic Ocean (d).

Imperfect volumetric sampling of the IC is a leading obstacle to further progress on several topics. For example, the scarcity of quasi-polar paths forces seismologists to make inferences on IC elastic properties (such as anisotropy, its nature and its lateral distribution in the IC) based on an incomplete seismological dataset. The fact that PKP(BC)−PKIKP and PKiKP−PKIKP datasets only sample the top

Figure 6.2 3D illustration of PKIKP ray paths traversing the IC from a view-point along the equatorial plane within the Earth. The yellow-orange globe in the centre of each image is the IC, and the map of the Earth's surface is projected onto the IC for orientation purposes. The ray paths are from existing datasets of PKP(BC)−PKIKP differential travel times collected through waveform correlation (Tkalčić et al., 2002; Leykam et al., 2010). The colours of the ray paths correspond to different values of travel time residuals: blue marks fast, white marks neutral, and red marks slow paths through the IC. Orange and yellow colours represent quasi-western and quasi-eastern hemispheres (qWH and qEH) of the IC, as defined by Tanaka and Hamaguchi (1997) (see Section 4.3.3). The IC as centred on the qEH and sampled by (a) quasi-polar PKIKP ray paths, defined by angle $\xi \leq 35°$ and (b) quasi-equatorial PKIKP ray paths, defined by angle $\xi \geq 35°$. The IC as centred on a transition between the qWH and qEH, sampled by (c) quasi-polar PKIKP ray paths, and (d) quasi-equatorial PKIKP ray paths. (A black and white version of this figure will appear in some formats. For the colour version, please refer to the plate section.)

Figure 6.3 3D illustration of PKIKP ray paths traversing the IC similar to Figure 6.2, but the IC is now viewed from a perspective within the Earth along the north-south axis. The IC as viewed from a point beneath the north pole, sampled by: (a) quasi-polar PKIKP ray paths, defined by an angle $\xi \leq 35°$ and (b) quasi-equatorial PKIKP ray paths, defined by an angle $\xi \geq 35°$. The IC as viewed from a point beneath the south pole, sampled by: (c) quasi-polar PKIKP ray paths, and (d) quasi-equatorial PKIKP ray paths. (A black and white version of this figure will appear in some formats. For the colour version, please refer to the plate section.)

part of the IC makes inferences on radial structure challenging. The shortage of crossing paths in the IC inhibits our ability to perform travel time tomography. The fact that global coverage is imperfect impacts our conclusions on attenuation properties in the IC and the relationship with isotropic and anisotropic velocity. On top of all of these challenges is the need to correct for seismic structure outside the IC before modelling the IC.

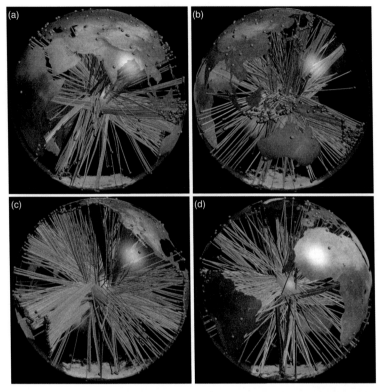

Figure 6.4 3D illustration of PKIKP ray paths traversing the IC from a viewpoint outside the Earth in the equatorial plane. The yellow globe in the centre of each image is the IC. The ray paths are from the existing datasets of PKP(BC)−PKIKP differential travel times collected through waveform correlation (Tkalčić et al., 2002; Leykam et al., 2010). Colours of ray paths correspond to different values of travel time residuals: blue marks fast, white marks neutral, and red marks slow paths through the IC. Green balls are station locations and red balls are event locations. The view is centred on (a) Indian Ocean, (b) southeast Asia, (c) Pacific Ocean and (d) Atlantic Ocean. (A black and white version of this figure will appear in some formats. For the colour version, please refer to the plate section.)

6.2 The Way Forward

6.2.1 How to Advance Seismological Studies of the IC

There are three proposed avenues for improving seismological data and tools as to progress our understanding of the IC:

a) increasing current coverage of the Earth's IC by global installations of continuously recording, modern broadband seismic instruments;
b) expanding the pool of seismic phases used in probing the IC;
c) revolutionising methods of analysing core-sensitive seismic phases.

How quickly the seismological community moves in the first direction depends on the resources available for future seismic deployments and success in communicating the importance of these installations to funding agencies. We will discuss in the following sections some examples of proposed deployments in the southern hemisphere. The second and the third points of potential improvement present observational and theoretical challenges, overcoming which will likely prove vital in the years to come.

6.2.2 Seismic Arrays and Stations in Remote Areas

Land

In Chapter 2 we saw how array seismology can enhance seismic signals coming from the deep Earth. In particular, we showed the benefits of using arrays with carefully chosen element configurations and argued that spiral arrays are optimal for detecting weak signals from the deep Earth. Such arrays also reduce side lobes in the array response function (see Section 2.4.2).

Recently three short aperture spiral arrays were deployed in Australia. Geoscience Australia installed the spiral Pilbara Seismic Array (PSAR) to enhance tsunami warning. SQspa, a spiral array in southern Queensland, and WAspa, a spiral array in Western Australia, were both installed by the ANU Seismology and Mathematical Geophysics group at RSES as a part of the Australian Research Council Discovery Project DP130101473 (Tkalčić, H., Kennett, B.L.N. and Tanaka, S.). Together with the WRA array, these arrays allow triangulation of deep Earth heterogeneity. Spiral arrays enhance detection capacity while keeping the number of elements and the cost to a minimum.

We also showed in Chapter 3 that regional arrays, in particular Hi-net in Japan due to its quality and unique geographic position, present powerful tools to probe the IC. Several recent studies have used Hi-net data to study various properties of the IC, e.g. attenuation (Iritani et al., 2010), isotropic velocity (Yee et al., 2014), and ICB topography (Tanaka and Tkalčić, 2015). Other seismic arrays such as the WOMBAT array (Australia) and USArray (United States) or their subarrays can be used to increase the signal-to-noise ratios of weak IC-sensitive seismic phases and form part of a multi-array for probing short-scale details.

Oceans

A significant increase in instrument coverage is unlikely due to the saturation of installations in developed countries. Thus the biggest benefit will likely come from the installation of ocean bottom stations positioned in boreholes reaching beneath the sediments, a.k.a. OBBS (ocean bottom borehole seismometers). Observations made by an OBBS sensor installed experimentally by the Japan Agency for

Marine Science and Technology (JAMSTEC) revealed PKP arrivals. PKP waves with dominant frequencies around 1 Hz are not normally visible on data from the typical ocean bottom seismometers that are resting on the ocean bottom or installed in shallow boreholes (S. Tanaka, personal communication, 2011) due to ambient noise. OBBS installations are technologically feasible but expensive; another option (though not significantly cheaper) is to utilise small oceanic islands in the southern hemisphere.

Figure 6.5 illustrates an example of a proposed deployment near the Macquarie archipelago in the mid-ocean Macquarie ridge (MR) complex at the boundary between the Indo-Australian and Pacific plates, approximately equidistant from New Zealand and Antarctica. The project would utilise both island-based and ocean bottom seismometers. This passive seismic experiment is capable of probing 3D structure of the oceanic crust, which should help answer fundamental questions about changes in crustal architecture throughout this part of the mid-ocean ridge. This will enable further understanding of the central part of the MR and its associated earthquakes, in particular, what physical conditions permitted the occurrence of the largest twentieth century strike-slip earthquake (23 May 1989, $M_w = 8.2$). Apart from the obvious benefits that such a seismological study would have in revealing the 3D structure of the MR oceanic crust, the configuration would allow for efficient amplification of seismic signals travelling through the deep Earth from earthquakes on the other side of the planet. The proposed experiment would help constrain regional Earth structure and improve the spatial coverage of the Earth's southern hemisphere, thereby making a noteworthy contribution to global imaging of Earth's deep interior.

Figure 6.6 illustrates the quality of the proposed arrays (MR X and MR SPIRAL arrays featured in Figure 6.5). The excellent resolution achieved for the proposed arrays can be appreciated if a comparison is made with the response of a short-aperture WRA array (Warramunga, Northern Territory), widely considered the top-performing International Monitoring System station. This comparison reveals that the proposed design has several outstanding features and exceeds the quality of the WRA response despite the fact that the sub-arrays contain a smaller number of elements than WRA (12 (spiral) and 18 (X) in comparison with WRA's 24 elements). Not only is the power concentrated in an area narrower than that seen for WRA, but the side-bands are also suppressed in comparison with WRA's response. This is due to the versatility of inner-station vectors, which is imposed by the experimental design (see Section 2.4.2).

Figure 6.7 illustrates that it would only take a few carefully planned station locations and state of the art instruments to significantly improve our capacity to capture PKP waves from the most prominent earthquakes occurring in the northern hemisphere. Frequent strong earthquakes would produce a large volume of

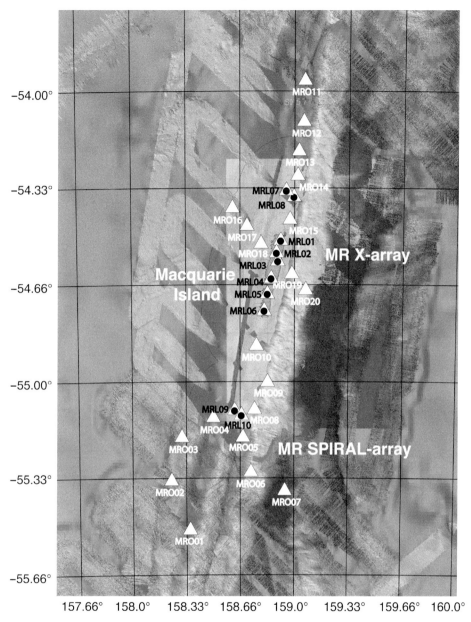

Figure 6.5 Location of the proposed deployment along the central portion of the Macquarie ridge (MR) in the vicinity of Macquarie Island. The available lineage of the bathymetry dataset (Spinoccia, Geoscience Australia, written communication) is shown in greyscale, ranging from about 70 m near the ridge to about 6300 m to the east (dark grey zone). The proposed locations for the ocean bottom and land (main island and islets) instruments are represented by triangles and the superimposed dots on triangles, respectively. Note an X-shaped sub-array in the northern section and a spiral-shaped sub-array in the southern section. MRO (triangles): ocean bottom instruments; MRL (triangles with a black dot): land (island and islets) instruments.

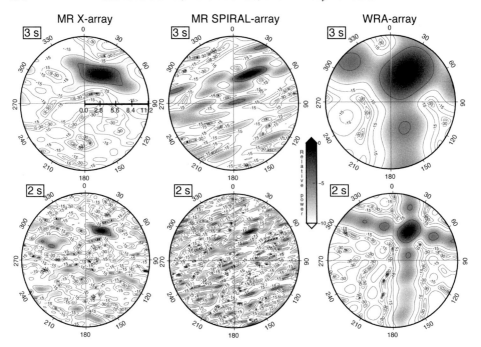

Figure 6.6 Theoretical array response functions calculated in the frequency domain for the proposed sub-array configurations (Figure 6.5): the letter x-shaped configuration comprising of 20 land and OBS elements (left column) and the three spiral arm-shaped configuration comprising of ten OBS elements (central column). The relative power in decibels is plotted as a function of horizontal slowness along the radial axis (in s°) with 0 in the centre and back azimuth in the clockwise direction forming an angle from geographic north. Array responses are shown in the slowness range 0–11.2s° for a monochromatic 3 s (0.33 Hz) wave (top row) and 2 s (0.5 Hz) wave (bottom row) entering the array domain at the slowness of 5.1s° and at the back azimuth of 30°. Theoretical array response of a short-aperture WRA array for the same monochromatic waves (right column).

PKP observations with unprecedented IC ray path coverage. For example, a borehole station in the South Indian Ocean on the Kerguelen Islands would capture earthquakes from the eastern part of the Aleutian Islands and from North America, most importantly including large events from Alaska, the Pacific Northwest region, and the coast of California. A borehole station in the South Pacific Ocean off the coast of Chile on the Humboldt Plane would capture most earthquakes from the central trans-Asian belt and the northwest coast of the Pacific Ocean, including Japan and the Sea of Okhotsk. Installing a borehole station a bit further west in the central part of the South Pacific would capture the Mediterranean earthquakes. Alaskan and Aleutian Islands' earthquakes would be captured by a borehole station near or on the South Sandwich Islands. Most of these newly captured PKP waves

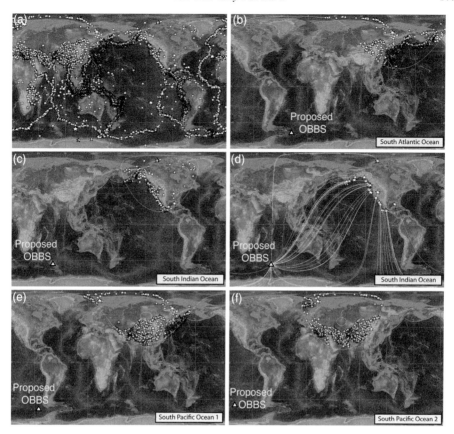

Figure 6.7 World seismicity from the ISC catalogue (white stars: shallow earthquakes, yellow stars: intermediate earthquakes, and red stars: deep earthquakes). (a) global earthquakes available in the ISC catalogue for the period 2000–2010. (b) a hypothetical station location in the South Atlantic Ocean near the South Sandwich Islands and all earthquakes in the epicentral distance range $145° \leq \xi \leq 175°$ that can potentially yield PKP observations. (c) the same as (b), but the hypothetical station location is in the South Indian Ocean near the Kerguelen Islands. (d) the same as (c), but only with earthquakes with $m_b \leq 5.5$ and including the corresponding great circle paths. (e) the same as (b), but the hypothetical station location is in the eastern part of South Pacific. (f) the same as (b), but the hypothetical station location is in the central part of South Pacific. Modified from Tkalčić (2015). (A black and white version of this figure will appear in some formats. For the colour version, please refer to the plate section.)

would traverse the IC quasi-parallel to the rotation axis of the Earth and would open new horizons in understanding IC velocity and anisotropic structure. Moreover, a borehole installation at the South Sandwich Islands would create a reverse ray path to the existing one, which would aid studies of IC rotational dynamics.

6.2.3 Exotic Seismic Phases

Sparse Observations

As noted above, apart from building observational infrastructure, another direction that should be taken in studying the IC is to search for little-used 'exotic' seismic phases that traverse the IC. For example, compressional waves that reach the CMB and ICB and then refract to the IC multiple times could in theory be used to probe the IC. These phases are rarely sought after or analysed due to the fact that they are highly attenuated in the IC (through scattering and viscoelastic attenuation), and are consequently more difficult to observe as their signal is buried in microseismic and event-generated noise. When they are observed, the seismological community is usually divided with regards to the validity of the observations.

In comparison to other research on the IC, an insignificant amount of time is invested in the detection of exotic phases. However, there have been a few recent studies reporting observations of exotic seismic phases, most notably those of PKIIKP and PKIKPPKIKP waves.

PKIKPPKIKP

Ray paths through the Earth of some exotic seismic phases are shown in Figure 2.3c. PKIKPPKIKP (a.k.a. P′P′df) waves are PKIKP waves that traverse the IC, reflect back from the internal side of the Earth's surface, and traverse the IC one more time before they reach the Earth's surface again. They are theoretically predicted and were observed in abundance during the Cold War era after nuclear explosions. Observations of their precursors, which are back-scattered from discontinuities in the upper mantle, have been used to discover upper mantle discontinuities. An example was shown in Section 2.4.2 (see Figure 2.12), and we will show another example in this section.

PKIKPPKIKP data should be considered complementary to PKIKP data and relevant for studies of isotropic and anisotropic structure and attenuation of the IC. Observations of antipodal PKIKP waves that sample near the Earth's centre are sparse, and non-existent in quasi-polar directions (see Figure 6.1). However, imagine that we can use PKIKPPKIKP waves instead of PKIKP waves, in particular those sampling the IC where raypath coverage is currently unavailable.

Complementing PKIKP with PKIKPPKIKP data is desirable because PKIKPP-KIKP waves are observed in a different angular distances range than PKIKP waves. In practice this means that we can utilise source-receiver paths that do not exist for PKIKP waves. To better illustrate this idea we have selected all events between April 2004 and April 2005 larger than magnitude 5.5 that can result in PKIKP data observed at a station in Berkeley, California (Figure 6.8a). We then identify events that are positioned on the globe as to yield PKIKPPKIKP data at the Berkeley

(a)

(b)

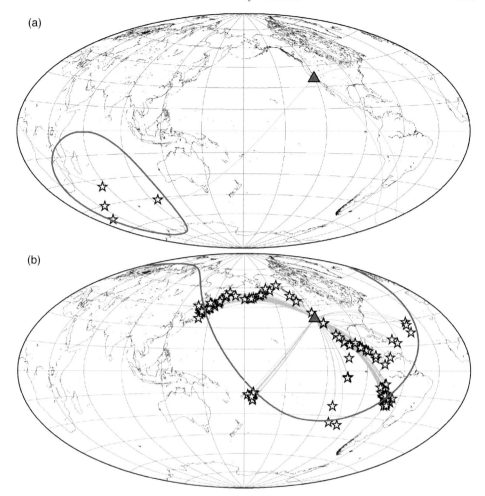

Figure 6.8 Illustration of the potential to probe the IC with PKIKPPKIKP waves using existing large earthquakes and permanent stations around the globe. (a) A station located in Berkeley, California, is shown by a triangle. Events with $m_b \geq 5.5$ and located in an epicentral distance range amenable to observing PKIKP waves suitable for differential travel time analysis ($145° \leq \Delta \leq 180°$) are shown by stars. (b) The same station and events with the same magnitude selection criterion, but now confined within the epicentral distance range that theoretically yields PKIKPPKIKP waves ($0° \leq \Delta \leq 90°$).

station (Figure 6.8b). Due to the existence of a large number of earthquakes near central and South America, there is a significant difference between the number of possible events associated with PKIKP and PKIKPPKIKP data recorded on that station.

The use of PKIKPPKIKP data would take advantage of the spatial co-existence of large earthquakes and seismic stations, a phenomenon that is quite common since more instruments are installed in earthquake-prone regions. If a station and the hypocentre of an earthquake are separated by less than 20°, we refer to the associated PKIKPPKIKP waves as 'podal', as in the opposite of 'antipodal'. The advantage of using podal PKIKPPKIKP waves is that they traverse the IC's centre twice. They are theoretically predicted, but observations have not been reported prior to the publications of Tkalčić and Flanagan (2004) and Tkalčić et al. (2006). These observations attest to a high quality factor (low attenuation) near the Earth's centre.

An example of a clear observation of near-podal PKIKPPKIKP waves is shown in Figure 6.9. A clear onset of PKIKPPKIKP waves in both the time and frequency domains is revealed by stacking the signals of two nuclear explosions from the late 1980s at the Nevada Test Site. The compressional waves propagated nearly twice through the Earth's centre. The observed absolute travel time of PKIKPPKIKP waves is close to that predicted by ak135 (Kennett et al., 1995). More observations of podal PKIKPPKIKP waves should be made before this dataset becomes a standard part of the IC-sensitive phases dataset available to researchers.

PKIIKP

PKIIKP waves are compressional waves whose leg traversing the IC is duplicated by the reflection from the lower side of the ICB (Figure 2.3c). At antipodal distances PKIIKP waves traverse a shallow part of the IC. Massé et al. (1974) reported clear arrivals of PKIIKP and a variety of other exotic phases at LASA following the large Novaya Zemlya nuclear event of 27 October 1973. They used the observed travel times (at an epicentral distance of 73°) in conjunction with PcP waves to estimate the average velocity in the outer half of the IC. Equally interesting are higher multiples of PKIKKIKP, denoted as PnKP waves (a.k.a. the 'whispering gallery' of phases), where n denotes the number of legs in the outer and inner cores. For example, P13KP was observed at the UC Berkeley Seismological Station during the era of nuclear tests in Novaya Zemlya (Bob Uhrhammer, personal communication, 1998).

PKIIKP waves are excellent companions to antipodal PKIKP waves. For example, they can serve as a reference when calculating differential travel times (Niu and Chen, 2008). This study found that waves traversing the IC in the quasi-equatorial plane are faster by about 1.8 s than those travelling in a direction that forms an angle of 28° with respect to the equatorial plane. This observation was interpreted as evidence for distinct anisotropy in the innermost 300 km of the IC, with the slow axis tilted by 45° with respect to the equatorial plane.

Potentially useful for studying UIC and lowermost OC structure (as well as temporal changes near the ICB) are observations of antipodal PKIKP data (and corresponding doublets). A recent waveform modelling study of antipodal arrivals

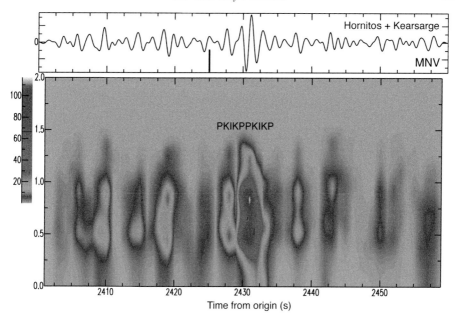

Figure 6.9 An observation of **PKIKPPKIKP** waves at station MINA, Nevada, after a zero-phase stack of the waveforms produced by two 150 kT nuclear explosions (Kearsarge, 17 August 1988 and Hornitos, 31 October 1989) (Tkalčić and Flanagan, 2004). The average epicentral distance from the explosions to the station is slightly less than 2°. A clear onset of PKIKPPKIKP is shown both in the time domain (seismogram on the top) and in the frequency domain (spectrogram at the bottom). The thick vertical line at 2425 s is the predicted travel time from ak135 (Kennett et al., 1995). The ray path associated with this observation samples the Earth about 50 km from its centre. (A black and white version of this figure will appear in some formats. For the colour version, please refer to the plate section.)

and their codas at TAM station in Northern Africa (Butler and Tsuboi, 2010) found that these waveforms cannot be explained by spherically symmetric Earth models such as PREM and instead require regional variations in velocity above and below the ICB.

Waszek and Deuss (2015) and Cormier (2015) recently reported new observations of PKIIKP waves. Waszek and Deuss (2015) stacked long period data and analysed the slowness of core-sensitive phases to identify possible recordings of PKIIKP waves. Cormier (2015) observed antipodal PKIIKP at the station TAM in Algeria (seismograms recorded at antipodal distances by the station TAM are featured in Figure 4.1). Waveform modelling suggests that these observations can be explained by a near-zero shear modulus in the qEH of the UIC within a layer 20–40 km thick in the region where PKIIKP waves sample the IC. According to Cormier (2015) this result, in combination with other seismological evidence (attenuation, scattering, and isotropic velocity), suggests that the qEH solidifies faster than the qWH and has a small grain size (10 m). This therefore agrees with the geodynamical models of Aubert et al. (2008) and Gubbins et al. (2011) and disagrees with the geodynamical models of Alboussière et al. (2010) and Monnereau et al. (2010).

Dense regional arrays will play a major role in the detection of more exotic phases associated with the IC, as they enable the construction of vespagrams (plots of relative slowness versus relative travel time) and fk stacks (see Section 2.9).

6.2.4 New Techniques to Analyse Waveform Data

Non-Linear Inversion to Measure PKP Travel Times

In the previous sections we discussed possible directions that could be taken to improve the body wave ray path coverage of the IC. But there is also the possibility of using existing waveforms that contain potentially usable PKIKP data that have previously been discarded as too noisy. A large part of the world's seismicity is comprised of shallow earthquakes that have been largely ignored in IC studies. Shallow earthquakes release PKIKP waves followed by depth phases of similar slowness. In the vicinity of PKP triplication (where PKIKP, PKP(BC), and PKP(AB) arrive simultaneously), the depth phases interfere with the main PKP phases and cause misidentification. In addition, most shallow, large earthquakes are strike-slip with radiation patterns that do not promote a strong emission of energy in the direction of PKIKP waves. Therefore, shallow earthquakes are often avoided and consequently excluded from analyses of high-quality records of core-sensitive waves. However, if the PKIKP data from these waveforms could be harvested, the potential uses would be numerous.

Garcia et al. (2004) and Garcia et al. (2006) developed a technique to utilise shallow earthquake data based on a theory put forward by Chevrot (2002). The method consists of selecting and preparing seismograms for a given earthquake and then running a simulated annealing waveform inversion of the body waves (SAWIB). Instead of directly picking PKP onsets on seismograms, the arrival times are obtained through non-linear inversion, with an option to account for the source-time function and depth phases. The synthesis of theoretical waveforms is based on the assumption that the PKIKP phase is an attenuated version of PKP(BC), and that the PKP(AB) phase is a Hilbert transform of the PKP(BC) phase.

An example of waveform fits after the SAWIB inversion is given for a deep earthquake in Fiji (Figure 6.10). An advantage of this approach is that multiple records on various recording stations from a single earthquake can be analysed. A computation of best-matching synthetic seismograms results in a less subjective approach to comparing waveforms and computing differential travel times, especially for complex records that would normally be discarded. However, the technique yields differential travel time data with a larger scatter than a carefully collected dataset where each waveform and a subsequent measurement is inspected by a seismologist.

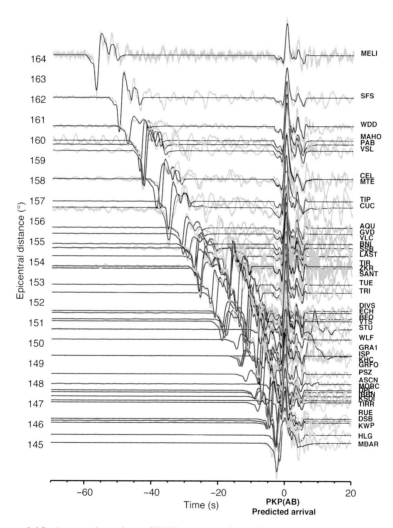

Figure 6.10 A record section of PKP waves at broadband stations available from IRIS DMC for an $m_b = 7.3$ earthquake in Fiji with a complex source-time function (grey lines). The analysed seismograms are in the epicentral distance range $145° \leq \Delta \leq 164°$. The records are aligned with respect to the predicted times of the PKP(AB) phase according to ak135 (Kennett et al., 1995). The synthetic seismograms computed by SAWIB are shown in black. After Tkalčić (2015).

Garcia et al. (2006) collected 4463 PKP traces from deep earthquakes (Figure 2.7b) and 7595 traces from shallow earthquakes (Figure 2.8e) using SAWIB. However, only about one-third of the shallow earthquakes yielded stable results owing to a poor signal-to-noise ratio of the PKIKP phase. The technique has been used *inter alia* to study attenuation in the shallow part of the IC (Iritani et al., 2010, 2014) and has been recently improved to account for the source time function (Garcia et al., 2013).

Seismic Interferometry

Seismic interferometry (e.g. Curtis et al., 2006; Schuster, 2009) is a promising, recently popular, and rapidly emerging method in deep-Earth global seismology. The method exploits pairs of seismograms (with both seismic noise and signal) and the similarity between them (computed by crosscorrelation or autocorrelation) to reconstruct the impulse response of the subsurface medium. The chosen seismogram pairs can be taken either from two different stations or from the same station. Studies initially focused on the shallow Earth (e.g. Snieder et al., 2002; Campillo and Paul, 2003; Shapiro et al., 2005), then gradually expanded to include the mantle transition zone (e.g. Poli et al., 2012), and finally imaging was performed on the global scale (e.g. Ruigrok et al., 2008; Boué et al., 2013; Nishida, 2013) and included the IC (e.g. Lin and Tsai, 2013; Lin et al., 2013; Wang et al., 2015).

The most extensive recent IC analysis has been performed by Huang et al. (2015), who analysed the autocorrelation of earthquake codas recorded across the USArray to detect $PKIKP^2$ and $PKIIKP^2$ body waves. $PKIKP^2$ waves are perfectly podal PKIKP waves, as discussed in Section 6.2.3. They travel through the Earth's centre, bounce back from the antipodal point at the Earth's surface, and travel through the Earth's centre again to return to the same station. Similarly, $PKIIKP^2$ are antipodal PKIIKP waves that travel back through the shallow part of the IC to the same station. Based on the travel time residuals between these two phases, they found strong and complex structural variability in the inner parts of the IC. Seismic interferometry already has numerous applications in deep Earth studies, and our understanding of IC structure and dynamics will continue to improve as refinements to the method are made.

Other Advances and Future Prospects

Unfortunately, full waveform modelling including 3D effects at frequencies of interest (above 1–2 Hz) is computationally expensive and still beyond reach for the purposes of studying the IC. Butler and Tsuboi (2010) were able to synthesise antipodal waveforms containing frequencies of up to 0.285 Hz (3.5 s) using the spectral element method (SEM) (Komatitsch et al., 2002). Waszek and Deuss (2015) used the same method to synthesise waveforms for their global search of exotic phases for frequencies up to 0.1 Hz (10 s). There are other less accurate, but consequently less computationally expensive, methods to synthesise seismograms; for example, Cormier (2015) used the full wave method of integration over ray parameter in a radially symmetric Earth (Rial and Cormier, 1980), and a superposition of Gaussian beams in a laterally-varying Earth (Červený, 1985).

Normal mode coupling theory has recently considered 'full coupling' of modes to be theoretically supported in addition to 'single' or 'group' coupling (Irving et al., 2009), which has led to both similarities and differences between various synthetic spectra of some IC sensitive modes in comparison with the observed spectra (Deuss et al., 2010; Deuss, 2014) that can be used to infer IC isotropic and anisotropic structure. There is often an apparent discrepancy in seismological observations when two very different observational tools, such as normal modes and body waves, are employed to investigate the same phenomenon. This often forces us to 'tune' our observational tools according to the central goal in order to reconcile measurements.

Recent advances in mathematical geophysics, in particular in the field of Bayesian inference (for a review, see Sambridge et al., 2013), present promising potential for IC rotation studies, and have resolved at least one existing discrepancy between normal modes and body waves (Tkalčić et al., 2013b,a). It is expected that mathematical geophysics will continue to play a significant role in the next decade, before full waveform modelling at higher frequencies of global phases sensitive to the IC becomes possible.

The resolution of various inconsistencies and differences between the models that guide our interpretation of IC properties and dynamic processes is gradually becoming within reach as better observational infrastructure develops. Deployments at strategic locations on both land and ocean bottom of short-aperture arrays similar to the Australian spiral arrays are now feasible. With the current global-scale expansion of seismic deployments (e.g. USArray in the USA, WOMBAT array in Australia, Hi-net array in Japan) and improvements in data quality, we are optimistic. Ocean bottom borehole seismometers are expensive, but their installation at a few strategic locations is a feasible step. The use of short, medium, and long aperture, simultaneously operating arrays should advance our capabilities to characterise IC structure. Improvements to ray path coverage and the capabilities of the rapidly developing field of seismic interferometry might open a new era in observational seismology and help to reveal the IC in a more comprehensive way than ever before.

Appendix A

Transmission/Reflection Coefficients for the Flat Inner Core Boundary

Consider an incident compressional (P) wave with amplitude A onto the horizontal and flat liquid–solid boundary, such as in Figure A.1. C is the amplitude of the reflected P wave (this corresponds to PKiK leg of PKiKP waves) and D is the amplitude of the refracted P wave (this corresponds to PKI leg of PKIKP waves). B is the amplitude of the refracted, vertically polarised, shear wave SV (this corresponds to PKJ leg of PKJKP waves). The angles are measured from the horizontal boundary to simplify the algebra in the transmission/reflection coefficient equations. α_1 and ρ_1 are the compressional velocity and density of the liquid core on the upper side of the boundary, whilst α_2, β_2, and ρ_2 are the compressional and shear velocities, and density of the solid core immediately beneath the boundary (note that these values could be taken from 1D models of the Earth). The x axis is positive upwards and z axis is positive eastwards.

The way the compressional energy is partitioned at this boundary can be expressed as the ratio of the amplitudes, which describes the amplitude of the transmitted wave to the solid core or reflected wave to the liquid core relative to an incident wave. We can refer to A/B and A/D as the transmission coefficients and C/A as the reflection coefficient. Our goal here is to express transmission and reflection coefficients as a function of ray geometry and known elastic properties at the ICB.

According to Snell's law:

$$\frac{\cos \varphi}{\alpha_1} = \frac{\cos \varphi'}{\alpha_1} = \frac{\cos \varphi''}{\alpha_2} = \frac{\cos \varphi'''}{\beta_2}. \tag{A.1}$$

Furthermore, the relations between compressional and shear wave velocities and the elastic parameters are:

$$\alpha = \sqrt{\frac{\lambda + 2\mu}{\rho}}; \beta = \sqrt{\frac{\mu}{\rho}}, \tag{A.2}$$

where λ and μ are Lamé constants of elasticity and ρ is the density.

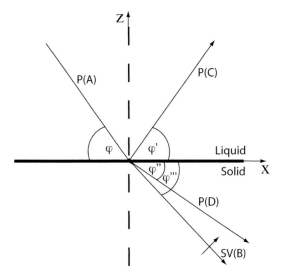

Figure A.1 Schematic representation of the incoming P wave and the components of the transmission/reflection system at the liquid–solid boundary in the
(x, z) plane, which can be applied to the inner core boundary (ICB) case. The
amplitudes of incoming, transmitted, and reflected waves are shown by capital
letters. For compressional waves, the displacement is parallel to the direction of
propagation (shown by arrows). The small arrow associated with the SV wave
indicates the orientation of the vector of ground displacement for shear waves.

For each wave, the wave number vector $\vec{k}(k_x, 0, k_z)$ is defined in the direction of
wave propagation and its absolute value equals the angular frequency ω divided by
the velocity (α or β). Each component of the wave number vector is a function of
the incident angle.

$$\text{Incident P wave: } k_x = \frac{\omega}{\alpha_1}\cos\varphi; \; k_z = -\frac{\omega}{\alpha_1}\sin\varphi;$$

$$\text{Reflected P wave: } k'_x = \frac{\omega}{\alpha_1}\cos\varphi'; \; k'_z = -\frac{\omega}{\alpha_1}\sin\varphi';$$

$$\text{Refracted P wave: } k''_x = \frac{\omega}{\alpha_2}\cos\varphi''; \; k_z = -\frac{\omega}{\alpha_2}\sin\varphi''; \tag{A.3}$$

$$\text{Refracted SV wave: } k'''_x = \frac{\omega}{\beta_2}\cos\varphi'''; \; k_z = -\frac{\omega}{\beta_1}\sin\varphi'''.$$

The ground motion vector $\vec{u}(u_x, 0, u_z)$ is defined in the direction of particle
motion. In the liquid core:

$$u_x = A\cos\varphi\, e^{ikx} + C\cos\varphi'\, e^{ik'x};$$

$$u_z = -A\sin\varphi\, e^{ikz} + C\sin\varphi'\, e^{ik'z}. \tag{A.4}$$

For the solid core:

$$u_x = B sin\varphi''' e^{ik'''x} + D cos\varphi'' e^{ik''x};$$

$$u_z = B cos\varphi''' e^{ik'''z} - D sin\varphi'' e^{ik'z}.$$

(A.5)

Note that the direction of SV motion is rotated by 90° from P motion using the right-hand co-ordinate system convention. Three boundary conditions are considered for the liquid–solid boundary: 1) normal displacements are continuous, 2) tangential stresses vanish, and 3) normal stresses are continuous.

The continuity of normal displacements requires:

$$- A sin\varphi e^{ikz} + C sin\varphi' e^{ik'z} = B cos\varphi''' e^{ik'''z} - D sin\varphi'' e^{ik''z}.$$

(A.6)

At $z = 0$, and considering Snell's law from Equation A.1, Equation A.6 becomes:

$$- (A + C) sin\varphi = B \frac{\beta_2}{\alpha_2} cos\varphi'' - D sin\varphi''.$$

(A.7)

Finally, dividing by A and rearranging, the following equation is derived:

$$\beta_2 \frac{B}{A} - \alpha_1 tan\varphi \frac{C}{A} - \alpha_2 tan\varphi'' \frac{D}{A} = -\alpha_1 tan\varphi,$$

(A.8)

where C/A is the reflection coefficient and D/A and B/A are refraction coefficients of P and SV waves, respectively.

The second boundary condition, according to which tangential stresses σ_{xz} vanish, requires:

$$\sigma_{xz} = 2\mu\varepsilon_{xz} = \mu \left(\frac{\partial u_x}{\partial z} + \frac{\partial u_z}{\partial x} \right) = 0,$$

(A.9)

where ε_{xz} is the tangential strain, μ is Lamé constant of elasticity (rigidity), $\partial u_x / \partial z$ is the partial derivative of the x component of ground displacement in the z direction, and $\partial u_z / \partial x$ is the partial derivative of the z component of ground displacement in the x direction. In liquid this condition is always satisfied, and in a solid, equations A.1, A.3, A.4, A.5, and A.9 are used to derive:

$$\frac{-B}{\beta_2} sin^2\varphi'' - \frac{D}{\alpha_2} sin\varphi'' cos\varphi'' + \frac{B}{\beta_2} cos^2\varphi''' - \frac{D}{\alpha_2} sin\varphi'' cos\varphi'' = 0.$$

(A.10)

Dividing by A and rearranging, we obtain the following equation:

$$\beta_2 \left(tan^2\varphi''' - 1 \right) \frac{B}{A} + 2\alpha_2 tan\varphi'' \frac{D}{A} = 0.$$

(A.11)

The third boundary condition requires the continuity of normal stresses σ_{zz} in liquids and solids:

$$- \lambda_1 (\varepsilon_{xx} + \varepsilon_{zz})_{liquid} = \lambda_2 (\varepsilon_{xx} + \varepsilon_{zz})_{solid} + 2\mu_2 (\varepsilon_{zz})_{solid},$$

(A.12)

where ε_{xx} and ε_{zz} are the normal strains, and μ and λ are Lamé's constants of elasticity. With the partial derivatives defining the normal strains, Equation A.12 becomes:

$$\lambda_1 \left(\frac{\partial u_x}{\partial x} + \frac{\partial u_z}{\partial z} \right)_{liquid} = \lambda_2 \left(\frac{\partial u_x}{\partial x} + \frac{\partial u_z}{\partial z} \right)_{solid} + 2\mu_2 \left(\frac{\partial u_z}{\partial z} \right)_{solid}. \quad (A.13)$$

After calculating partial derivatives and using Snell's law, Equation A.13 leads to:

$$\lambda_1 \left(\frac{A}{\alpha_1} + \frac{C}{\alpha_1} \right) = \lambda_2 \frac{D}{\alpha_2} + 2\mu_2 \left(\frac{D}{\alpha_2} sin^2\varphi'' - \frac{B}{\beta_2} sin\varphi''' cos\varphi''' \right). \quad (A.14)$$

Now, using Equation A.2 and substituting expressions for ρ, Equation A.14 can be written as:

$$\rho_! \alpha_1 \alpha_2 (A + C) = \rho_2 \alpha_2^2 D - 2\rho_2 \beta_2^2 D cos^2\varphi'' - 2\rho_2 \alpha_2 \beta_2 B sin\varphi''' cos\varphi''', \quad (A.15)$$

and with a further rearrangement:

$$2\frac{\beta_2{}^3}{\alpha_1{}^2}tan\varphi''' \frac{B}{A} + \alpha_1 \frac{\rho_1}{\rho_2}sec^2\varphi \frac{C}{A} - \alpha_2 \frac{\beta_2^2}{\alpha_1^2}(tan^2\varphi''' - 1) \frac{D}{A} = -\frac{\rho_1}{\rho_2}\alpha_1 sec^2\varphi. \quad (A.16)$$

Thus, a system of three equations with three unknown transmission/reflection coefficients for a liquid–solid boundary is derived, and it can be written as:

$$\begin{bmatrix} \beta_2(tan^2\varphi''' - 1) & 0 & 2\alpha_2 tan\varphi'' \\ \beta_2 & -\alpha_1 tan\varphi & -\alpha_2 tan\varphi'' \\ 2\frac{\beta_2{}^3}{\alpha_1{}^2}tan\varphi''' & \alpha_1\frac{\rho_1}{\rho_2}sec^2\varphi & -\alpha_2\frac{\beta_2^2}{\alpha_1^2}(tan^2\varphi''' - 1) \end{bmatrix} \begin{bmatrix} B/A \\ C/A \\ D/A \end{bmatrix} = \begin{bmatrix} 0 \\ -\alpha_1 tan\varphi \\ \frac{\rho_1}{\rho_2}\alpha_1 sec^2\varphi \end{bmatrix}. \quad (A.17)$$

In the above expression, we can see that the density ratio (liquid to solid) is explicit and a mathematical convenience that enables the estimation of the density ratio if the ray geometry and other elastic parameters are known.

Appendix B

The Angle Between PKIKP Waves and the Rotation Axis of the Earth

Here we will derive a formula for ξ, the angle between PKIKP ray in the inner core (IC) and the rotation axis of the Earth. This angle measures the directional dependence of PKIKP wave travel times through the IC.

We will use vector calculus in Cartesian co-ordinates, with the centre of the co-ordinate system at the centre of the Earth and the z-axis parallel to the Earth's rotation axis. Let us assume a PKIKP source at location $S(\theta_1, \phi_1)$ and a receiver at location $R(\theta_2, \phi_2)$, where θ is colatitude ($90° -$ latitude), ϕ is longitude, and the radius of the Earth is assumed to be equal to 1. The components of the position vectors S and R in Cartesian co-ordinates are given by

$$
\begin{aligned}
S_x &= \sin\theta_1 \cos\phi_1, \\
S_y &= \sin\theta_1 \sin\phi_1, \\
S_z &= \cos\theta_1, \\
R_x &= \sin\theta_2 \cos\phi_2, \\
R_y &= \sin\theta_2 \sin\phi_2, \\
R_z &= \cos\theta_2.
\end{aligned}
\tag{B.1}
$$

We will use an approximation that, regardless of their true depth, the source and receiver are both at the surface, as this will not introduce significant errors even for deep earthquakes. It then follows that, in a spherically symmetric Earth, the PKIKP ray direction at the bottoming point is parallel to the vector $\vec{L} = (L_x, L_y, L_z)$ pointing from the source to the receiver, whose components are proportional to

$$
\begin{aligned}
L_x &= \sin\theta_1 \cos\phi_1 - \sin\theta_2 \cos\phi_2, \\
L_y &= \sin\theta_1 \sin\phi_1 - \sin\theta_2 \sin\phi_2, \\
L_z &= \cos\theta_1 - \cos\theta_2.
\end{aligned}
\tag{B.2}
$$

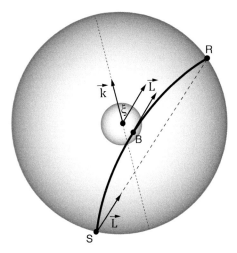

Figure B.1 Cross-section through the Earth and a PKIKP ray path. S is the source, R is the receiver, and B is the bottoming point.

Let $\vec{k} = (0, 0, 1)$ be a unit vector parallel to the rotation axis of the Earth. By vector translation, the angle ξ can then be calculated as a dot product

$$\cos \xi = \frac{\vec{L} \cdot \vec{k}}{|L||k|} = \frac{L_x \cdot 0 + L_y \cdot 0 + L_z \cdot 1}{\sqrt{L_x^2 + L_y^2 + L_z^2}\sqrt{1}}. \tag{B.3}$$

After some algebra, B.3 becomes

$$\cos \xi = \frac{\cos \theta_1 - \cos \theta_2}{\sqrt{2 - 2\cos \theta_1 \cos \theta_2 - 2\sin \theta_1 \sin \theta_2 \cos(\phi_1 - \phi_2)}}. \tag{B.4}$$

Appendix C

P-Wave Velocity in a Transversely Isotropic Inner Core

C.1 Expression for P-Wave Velocity as a Function of $cos^2\xi$

Here we will derive an expression for compressional (P wave) velocity, v_p, in a transversely isotropic IC as a function of $cos^2\xi$, where ξ is the angle between the ray and the rotation axis of the Earth. This expression has been repeatedly used in literature to compute velocity and travel time perturbations of PKIKP waves in the IC as a function of the angle between the PKIKP waves and the rotation axis of the Earth.

We start with the expression for P-wave velocity in a transversely isotropic medium:

$$v_p^2\rho = Asin^4\xi + Ccos^4\xi + 2(2L + F)sin^2\xi cos^2\xi, \tag{C.1}$$

where ρ is density, ξ is the angle between the P wave and the axis of symmetry, and A, C, L, and F are the elastic constants (Love, 1927). For a purely isotropic case, $A = C = 2L + F$.

Under this condition, the Equation C.1 simplifies as follows:

$$v_p^2\rho = Asin^4\xi + Acos^4\xi + 2Asin^2\xi cos^2\xi$$
$$= A(sin^2\xi + cos^2\xi)^2$$
$$= A.$$

Thus, isotropic P-wave velocity is given by:

$$v_p = \sqrt{\frac{A}{\rho}}. \tag{C.2}$$

Since isotropic P-wave velocity relates to the Lamé constants by $v_p = \sqrt{\frac{\lambda+2v}{\rho}}$, it follows that $A = \lambda + 2v$.

For a 'weakly anisotropic' case, we can consider a perturbation from an isotropic case (Morelli et al., 1986), with a reference velocity v_p defined in C.2 (termed 'equatorial' velocity). With these assumptions and convenient substitutions for the elastic constants, it is possible to derive a useful expression of P-wave velocity as a function of only two independent parameters, ϵ and σ.

The substitutions are as follows:

$$C = (1 + 2\epsilon)A,$$

$$2L + F = (1 + \sigma)\sqrt{AC}.$$

We can express these two substitutions in terms of the elastic constant A and substitute into Equation C.1. This leads to:

$$v_p^2 = \frac{A}{q}[sin^4\xi + (1 + 2\epsilon)cos^4\xi + 2(1 + \sigma)\sqrt{1 + 2\epsilon}\ sin^2\xi cos^2\xi]. \quad (C.3)$$

Now, recall Binomial series

$$(1 + 2\epsilon)^{1/2} \approx 1 + \frac{1}{2}(2\epsilon) + \ldots \approx 1 + \epsilon,$$

and that for weak anisotropy, we can neglect squares and cross terms of ϵ and σ. Thus, from C.3, we get

$$v_p^2 = \frac{A}{q}[sin^4\xi + cos^4\xi + 2\epsilon\ cos^4\xi + 2sin^2\xi cos^2\xi + 2\epsilon sin^2\xi cos^2\xi + 2\sigma sin^2\xi cos^2\xi].$$

After rewriting the last term as

$$sin^2\xi cos^2\xi = sin^2\xi cos^4\xi + cos^2\xi sin^4\xi,$$

we finally get

$$v_p^2 = \frac{A}{q}[sin^4\xi + cos^4\xi + 2sin^2\xi cos^2\xi + 2\epsilon\ cos^4\xi + 2\epsilon\ sin^2\xi cos^2\xi$$
$$+ 2\sigma sin^2\xi cos^2\xi + 2\sigma sin^4\xi cos^2\xi]. \quad (C.4)$$

When all squares and cross terms of ϵ and σ are neglected, the term in parentheses is equivalent to:

$$[sin^2\xi + cos^2\xi + \epsilon\ cos^2\xi + \sigma sin^2\xi cos^2\xi]^2.$$

Taking a square root of C.4, we obtain:

$$v_p = \sqrt{\frac{A}{\rho}}(1 + \epsilon\ cos^2\xi + \sigma sin^2\xi cos^2\xi). \quad (C.5)$$

Considering the isotropic velocity defined in C.2 as a reference velocity, we can express the P-wave velocity perturbation from C.5 as:

$$\frac{\delta v_p}{v_p} = \epsilon\ cos^2\xi + \sigma sin^2\xi cos^2\xi. \quad (C.6)$$

Appendix D

Transdimensional Bayesian Inversion

Within an optimisation framework, the number, shape, and size of basis functions used to parameterise a model space are typically decided in advance. Parameterisation examples include the use of blocks or spherical harmonics in seismic tomography, or the layers in receiver functions, to parameterise Earth structure. A Bayesian inversion, however, is an inference process in which a collection of models are sought instead of a single 'best fit' model. In its transdimensional extension, many models of a variable number of model parameters (model dimensions) are sampled. Transdimensional sampling eliminates the need for a subjective, a priori choice about the number of basis functions (model dimension) and makes that number an unknown itself.

Bayes' theorem, expressed in Equation 5.8, can be written with a fixed total number of unknowns, k:

$$p(\mathbf{m}|\mathbf{d}, k) \propto p(\mathbf{d}|\mathbf{m}, k) p(\mathbf{m}|k), \tag{D.1}$$

where $p(\mathbf{m}|\mathbf{d}, k)$ is a posterior distribution of fixed k. Sambridge et al. (2006) show that the above expression can be formulated in a transdimensional framework as

$$p(\mathbf{m}, k|\mathbf{d}) \propto p(\mathbf{d}|\mathbf{m}, k) p(\mathbf{m}|k) p(k), \tag{D.2}$$

where k is now a variable which has its own prior probability density function, $p(k)$. $p(\mathbf{m}, k|\mathbf{d})$ is a posterior distribution with a variable k. Each model is sampled based on the information provided by the prior distributions $p(\mathbf{m}|k)$ and $p(k)$, and the data, $\mathbf{d}$, with a corresponding likelihood function, $p(\mathbf{d}|\mathbf{m}, k)$. The transdimensional likelihood function in our regression problem can be expressed as

$$p(\mathbf{d}|\mathbf{m}, k) = \frac{1}{\prod_{i-1}^{N}\left(\sqrt{2\pi\sigma_i^2}\right)} \exp\left[\sum_{i-1}^{N} \frac{-(g(\mathbf{m})_i - \mathbf{d}_i)^2}{2(\sigma_i)^2}\right], \tag{D.3}$$

where N is the number of data points, $\mathbf{d}_i$ is the ith datum, $g(\mathbf{m})$ is the ith datum estimated from a given model, $\mathbf{m}$, and σ_i is the standard deviation of an assumed random Gaussian noise for measurement i.

'Hierarchical' sampling does not fix σ as in Equation D.3, but leaves it as a free parameter with its own prior. The data themselves are thus used to constrain the level of data uncertainty during the inversion, which in turn controls the degree of complexity of the solution. Here the data uncertainty refers to the inability to fit real-world observations, which may be due to measurement errors, inaccuracies in the forward theory or inappropriate model parameterisation. Regardless of whether the data variance is allowed to vary, this type of transdimensional sampling approach is 'naturally parsimonious' in that models with more unknowns are not necessarily favoured over those with fewer unknowns, even though a well-fitting model with more variables will usually better predict the data. Simple parameterisations are favoured over the more complex provided they are consistent with the data. More details about transdimensional inversion can be found in a review by Sambridge et al. (2013).

In our regression problem (Section 5.4.1), we seek a differential travel time slope, $\frac{\delta t}{\Delta T}$ (see Equation 5.5). The time interval is divided into a variable number of Voronoi partitions defined by a set of nuclei, whose number and position along the time axis are variables in the algorithm. The term *Voronoi partition* refers to a region that encloses all points in space that are closer to the partition's Voronoi nucleus than to any other nucleus. Thus, the shape and size of the basis functions are implicit. Alternatively, the model space can be partitioned using more explicit forms of basis functions, e.g. B-splines (Tkalčić et al., 2013b).

References

Aki, K., and Richards, P. G. 2002. *Quantitative Seismology: Theory and Methods*, 2nd edn. Sausalito, CA: University Science Books.

Alboussière, T., Renaud, D., and Mickaël, M. 2010. Melting-induced stratification above the Earth's inner core due to convective translation. *Nature*, **466**(June), 744–747.

Andrews, J., Deuss, A., and Woodhouse, J. 2006. Coupled normal-mode sensitivity to inner-core shear velocity and attenuation. *Geophysical Journal International*, **167**(1), 204–212.

Antonangeli, D., Merkel, S., and Farber, D. L. 2006. Elastic anisotropy in HCP metals at high pressure and the sound wave anisotropy of the Earth's inner core. *Geophysical Research Letters*, **33**(24).

Attanayake, J., Cormier, V. F, and de Silva, S. M. 2014. Uppermost inner core seismic structure–new insights from body waveform inversion. *Earth and Planetary Science Letters*, **385**, 49–58.

Aubert, J., Amit, H., Hulot, G., and Olson, P. 2008. Thermochemical flows couple the Earth's inner core growth to mantle heterogeneity. *Nature*, **454**(August), 758–761.

Aurnou, J., and Olson, P. 2000. Control of inner core rotation by electromagnetic, gravitational and mechanical torques. *Physics of the Earth and Planetary Interiors*, **117**(1–4), 111–121.

Bashō, M. 1694. *Oku no Hosomichi (Narrow Road to the Interior)*.

Bayes, T. 1763. An essay towards solving a problem in the doctrine of chances. *Philosophical Transactions*, **53**, 370–418.

Beghein, C., and Trampert, J. 2003. Robust normal mode constraints on inner-core anisotropy from model space search. *Science*, **299**(5606), 552–555.

Belonoshko, A. B., Skorodumova, N. V., Davis, S., Osiptsov, A. N., Rosengren, A., and Johansson, B. 2007. Origin of the low rigidity of the Earth's inner core. *Science*, **316**(5831), 1603–1605.

Belonoshko, A. B., Skorodumova, N. V., Rosengren, A., and Johansson, B. 2008. Elastic anisotropy of Earth's inner core. *Science*, **319**(5864), 797–800.

Bergman, M. I. 1997. Measurements of electric anisotropy due to solidification texturing and the implications for the Earth's inner core. *Nature*, **389**(6646), 60–63.

Bergman, M. I. 1998. Estimates of the Earth's inner core grain size. *Geophysical Research Letters*, **25**(10), 1593–1596.

Bergman, M. I. 2003. *Earth's Core: Dynamics, Structure, Rotation*, 1st edn. American Geophysical Union Geodynamics Series, vol. 31. Washington, DC: American

Geophysical Union Books Board. Chap. Solidification of the Earth's core, pages 105–127.

Bergman, M. I., and Fearn, D. R. 1994. Chimneys on the Earth's inner-outer core boundary. *Geophysical Research Letters*, 477–480.

Bergman, M. I., Cole, D. M., and Jones, J. R. 2002. Preferred crystal orientations due to melt convection during directional solidification. *Journal of Geophysical Research*, **107**.

Birch, A. F. 1940. The alpha-gamma transformation of iron at high pressures, and the problem of the earth's magnetism. *American Journal of Science*, **238**(3), 192–211.

Bloxham, J., and Gubbins, D. 1987. Morphology of the geomagnetic field and implications for the geodynamo. *Nature*, **325**(6104), 509–511.

Bloxham, J., Zatman, S., and Dumberry, M. 2002. The origin of geomagnetic jerks. *Nature*, **420**(November), 65–68.

Bolt, B. A. 1962. Gutenberg's early PKP observations. *Nature*, **196**, 122–124.

Bolt, B. A. 1972. The density distribution near the base of the mantle and near the Earth's core. *Physics of the Earth and Planetary Interiors*, **5**, 302–311.

Bolt, B. A., and Qamar, A. 1970. Upper bound to the density jump at the boundary of the Earth's inner core. *Nature*, **228**, 148–150.

Boué, P., Poli, P., Campillo, M., Pedersen, H., Briand, X., and Roux, P. 2013. Teleseismic correlations of ambient seismic noise for deep global imaging of the Earth. *Geophysical Journal International*, **194**, 844–848.

Braginsky, S. I. 1963. Structure of the F-layer and reasons for convection in the Earth's core. *Doklady Akad. Nauk SSSR*, **149**, 8–10.

Bréger, L., Romanowicz, B., and Tkalčić, H. 1999. PKP(BC-DF) Travel time residuals and short scale heterogeneity in the deep Earth. *Geophysical Research Letters*, **26**(20), 3169–3172.

Bréger, L., Tkalčić, H., and Romanowicz, B. 2000. The effect of D″ on PKP(AB-DF) travel time residuals and possible implications for inner core structure. *Earth and Planetary Science Letters*, **175**, 133–143.

Brush, S. G. 1980. Discovery of the Earth's core. *Am. J. Phys.*, **48**(9), 705–724.

Buchbinder, G. G. R. 1971. A velocity structure of the Earth's core. *Bulletin of the Seismological Society of America*, **61**(2), 429–456.

Buchbinder, G. G. R., Wright, C., and Poupinet, G. 1973. Observations of PKiKP at distances less than 110°. *Bulletin of the Seismological Society of America*, **63**(5), 1699–1707.

Buffett, B. A., and Wenk, H. R. 2001. Texturing of the Earth's inner core by Maxwell stresses. *Nature*, **413**(6851), 60–63.

Buffett, B. A. 1997. Geodynamic estimates of the viscosity of the Earth's inner core. *Nature*, **388**(August), 571–573.

Buffett, B. A., and Creager, K. C. 1999. A comparison of geodetic and seismic estimates of inner-core rotation. *Geophysical Research Letters*, **26**(10), 1509–1512.

Buffett, B. A., Huppert, H. E., Lister, J. R., and Woods, A. W. 1996. On the thermal evolution of the Earth's core. *Journal of Geophysical Research: Solid Earth*, **101**(B4), 7989–8006.

Buland, R., and Gilbert, F. 1978. Improved resolution of complex eigenfrequencies in analytically continued seismic spectra. *Geophysical Journal of the Royal Astronomical Society*, **52**(3), 457–470.

Bullen, K. E. 1946. A hypothesis on compressibility at pressures of the order of a million atmospheres. *Nature*, **157**(March), 405.

Butler, R., and Tsuboi, S. 2010. Antipodal seismic observations of temporal and global variation at Earth's inner-outer core boundary. *Geophysical Research Letters*, **37**(11).

Calvet, M., and Margerin, L. 2008. Constraints on grain size and stable iron phases in the uppermost inner core from multiple scattering modeling of seismic velocity and attenuation. *Earth and Planetary Science Letters*, **267**(March), 200–212.

Calvet, M., and Margerin, L. 2012. Velocity and attenuation of scalar and elastic waves in random media: A spectral function approach. *The Journal of the Acoustial Society of America*, **231**(3), 1843–1862.

Calvet, M., Chevrot, S., and Souriau, A. 2006. P-wave propagation in transversely isotropic media II. Application to inner core anisotropy: Effects of data averaging, parametrization and a priori information. *Physics of the Earth and Planetary Interiors*, **156**, 21–40.

Campillo, M., and Paul, A. 2003. Long-range correlations in the diffuse seismic coda. *Science*, **299**, 547–549.

Cao, A., and Romanowicz, B. 2004. Hemispherical transition of seismic attenuation at the top of the Earth's inner core. *Earth and Planetary Science Letters*, **228**(December), 243–253.

Cao, A., Romanowicz, B., and Takeuchi, N. 2005. An observation of PKJKP: Inferences on inner core shear properties. *Science*, **308**(June), 1453–1455.

Cao, A., Masson, Y., and Romanowicz, B. 2007. Short wavelength topography on the inner-core boundary. *Proceedings of the National Academy of Science*, **104**(January), 31–35.

Cao, A., and Romanowicz, B. 2004. Constraints on density and shear velocity contrasts at the inner core boundary. *Geophysical Journal International*, **157**(3), 1146–1151.

Cavendish, H. 1798. Experiments to determine the density of the Earth. By Henry Cavendish, Esq. F. R. S. and A. S. *Philosophical Transactions of the Royal Society of London*, **88**, 469–526.

Červený, V. 1985. Gaussian beam synthetic seismograms. *Journal of Geophysics*, **58**, 42–72.

Červený, V. 2005. *Seismic Ray Theory*, 1st edn. Cambridge University Press.

Chapman, C. H. 2004. *Fundamentals of Seismic Wave Propagation*. Cambridge University Press.

Chevrot, S. 2002. Optimal measurement of relative and absolute delay times by simulated annealing. *Geophysical Journal International*, **151**(1), 164–171.

Choy, G. L., and Cormier, V. F. 1983. The structure of the inner core inferred from short-period and broadband GDSN data. *Geophysical Journal of the Royal Astronomical Society*, **72**, 1–21.

Collier, J. D., and Helffrich, G. 2001. Estimate of inner core rotation rate from United Kingdom regional seismic network data and consequences for inner core dynamical behaviour. *Earth and Planetary Science Letters*, **193**(3–4), 523 – 537.

Cormier, V. F. 1981. Short-period PKP phases and the anelastic mechanism of the inner core. *Physics of the Earth and Planetary Interiors*, **24**(4), 291–301.

Cormier, V. F. 2007. Texture of the uppermost inner core from forward- and back-scattered seismic waves. *Earth and Planetary Science Letters*, **258**(3–4), 442 – 453.

Cormier, V. F. 2009. A glassy lowermost outer core. *Geophysical Journal International*, **179**(1), 374–380.

Cormier, V. F. 2011. *Encyclopedia of Earth Sciences Series.* Springer Netherlands. Pages 1279–1290.

Cormier, V. F. 2015. Detection of inner core solidification from observations of antipodal PKIIKP. *Geophysical Research Letters*, **42**, 7459–7466.

Cormier, V. F., and Attanayake, J. 2013. Earth's solid inner core: Seismic implications of freezing and melting. *Journal of Earth Science*, **24**(5), 683–698.

Cormier, V. F., and Choy, G. L. 1986. A search for lateral heterogeneity in the inner core from differential travel times near PKP-D and PKP-C. *Geophysical Research Letters*, **13**(13), 1553–1556.

Cormier, V. F., and Li, X. 2002. Frequency-dependent seismic attenuation in the inner core 2. A scattering and fabric interpretation. *Journal of Geophysical Research (Solid Earth)*, **107**(December), 2362.

Cormier, V. F., and Richards, P. G. 1976. Comments on 'The Damping of Core Waves' by Anatholy Qamar and Alfredo Eisenberg. *Journal of Geophysical Research: Solid Earth*, **81**, 3,066–3,068.

Cormier, V. F., and Stroujkova, A. 2005. Waveform search for the innermost inner core. *Earth and Planetary Science Letters*, **236**(July), 96–105.

Cormier, V. F., Xu, L., and Choy, G. L. 1998. Seismic attenuation of the inner core: Viscoelastic or stratigraphic? *Geophysical Research Letters*, **25**, 4019–4022.

Cormier, V. F., Attanayake, J., and He, K. 2011. Inner core freezing and melting: Constraints from seismic body waves. *Physics of the Earth and Planetary Interiors*, **188**(3), 163–172.

Creager, K. C. 1997. Inner core rotation rate from small-scale heterogeneity and time-varying travel times. *Science*, **278**, 1284–1288.

Creager, K. C. 1992. Anisotropy of the inner core from differential travel times of the phases PKP and PKIKP. *Nature*, **356**(March), 309–314.

Creager, K. C. 1999. Large-scale variations in inner core anisotropy. *Journal of Geophysical Research*, **104**(B10), 23, 127–123, 139.

Cummins, P., and Johnson, L. R. 1988. Short-period body wave constraints on properties of the Earth's inner core boundary. *Journal of Geophysical Research: Solid Earth*, **93**, 9058–9074.

Curtis, A., Gerstoft, P., Sato, H., Snieder, R., and Wapenaar, K. 2006. Seismic interferometry turning noise into signal. *The Leading Edge*, **25**, 1082–1092.

Dahlen, F. A., and Tromp, J. 1998. *Theoretical Global Seismology*. Princeton University Press.

Dai, Z., Wang, W., and Wen, L. 2012. Irregular topography at the Earth's inner core boundary. *Proceedings of the National Academy of Sciences*, **109**(20), 7654–7658.

Deguen, R. 2012. Structure and dynamics of Earth's inner core. *Earth and Planetary Science Letters*, **333**, 211.

Deguen, R., Alboussière, T., and Brito, D. 2007. On the existence and structure of a mush at the inner core boundary of the Earth. *Physics of the Earth and Planetary Interiors*, **164**(September), 36–49.

Deguen, R., and Cardin, P. 2011. Thermochemical convection in Earth's inner core. *Geophysical Journal International*, **187**, 1101–1118.

Deuss, A. 2014. Heterogeneity and anisotropy of Earth's inner core. *Annual Review of Earth and Planetary Sciences*, **42**(1), 103–126.

Deuss, A., Irving, J. C. E., and Woodhouse, J. H. 2010. Regional variation of inner core anisotropy from seismic normal mode observations. *Science*, **328**(5981), 1018–1020.

Deuss, A., Woodhouse, J. H., Paulssen, H., and Trampert, J. 2000. The observation of inner core shear waves. *Geophysical Journal International*, **142**(1), 67–73.

Doornbos, D. J. 1974. The anelasticity of the inner core. *Geophysical Journal of the Royal Astronomical Society*, **38**(2), 397–415.

Doornbos, D. J. 1988. *Seismological Algorithms: Computational Methods and Computer Programs*. Academic Press.

Doornbos, D. J., and Husebye, E. 1972. Array analysis of PKP phases and their precursors. *Physics of the Earth and Planetary Interiors*, **5**, 387–399.

Dubrovinsky, L., Dubrovinskaia, N., Narygina, O., Kantor, I., Kuznetzov, A., Prakapenka, V. B., Vitos, L., Johansson, B., Mikhaylushkin, A. S., Simak, S. I., and Abrikosov, I. A. 2007. Body-centered cubic iron-nickel alloy in Earth's core. *Science*, **316**(5833), 1880–1883.

Dumberry, M. 2007. Geodynamic constraints on the steady and time-dependent inner core axial rotation. *Geophysical Journal International*, **170**(2), 886–895.

Dumberry, M., and Mound, J. 2010. Inner core-mantle gravitational locking and the super-rotation of the inner core. *Geophysical Journal International*, **181**(2), 806–817.

Durek, J. J., and Romanowicz, B. 1999. Inner core anisotropy inferred by direct inversion of normal mode spectra. *Geophysical Journal International*, **139**(3), 599–622.

Dziewoński, A. M. 1984. Mapping the lower mantle: Determination of lateral heterogeneity in P velocity up to degree and order 6. *Journal of Geophysical Research: Solid Earth*, **89**(B7), 5929–5952.

Dziewoński, A. M., and Anderson, D. L. 1981. Preliminary reference Earth model. *Physics of the Earth and Planetary Interiors*, **25**, 297–356.

Dziewoński, A. M., and Gilbert, F. 1971. Solidity of the inner core of the Earth inferred from normal mode observations. *Nature*, **234**(December), 465–466.

Engdahl, E. R., Flinn, E. A., and Massé, R. P. 1974. Differential PKiKP travel times and the radius of the inner core. *Geophysical Journal of the Royal Astronomical Society*, **39**(3), 457–463.

Engdahl, E. R., Flinn, E. A., and Romney, C. F. 1970. Seismic waves reflected from the Earth's inner core. *Nature*, **228**(5274), 852–853.

Engdahl, E. R., Van der Hilst, R. D., and Buland, R. P. 1998. Global teleseismic earthquake relocation with improved travel times and procedures for depth determination. *Bulletin of the Seismological Society of America*, **88**, 722–743.

Fearn, D. R., Loper, D. E., and Roberts, P. H. 1981. Structure of the Earth's inner core. *Nature*, **292**(5820), 232–233.

Fowler, C. M. R. 2005. *The Solid Earth: An Introduction to Global Geophysics*, 2nd edn. Cambridge University Press.

Fukao, Y., and Suda, N. 1989. Core modes of the Earth's free oscillations and structure of the inner core. *Geophysical Research Letters*, **16**(5), 401–404.

Gans, R. F. 1972. Viscosity of the Earth's core. *Journal of Geophysical Research*, **77**(2), 360–366.

Garcia, R., Chevrot, S., and Weber, M. 2004. Nonlinear waveform and delay time analysis of triplicated core phases. *Geophysical Journal International*, **109**.

Garcia, R., Tkalčić, H., and Chevrot, S. 2006. A new global PKP data set to study Earth's core and deep mantle. *Physics of the Earth and Planetary Interiors*, **159**(1–2), 15–31.

Garcia, R. F. 2002. Constraints on upper inner core structure from waveform inversion of core phases. *Geophysical Journal International*, **150**, 651–664.

Garcia, R. F., Schardong, L., and Chevrot, S. 2013. A nonlinear method to estimate source parameters, amplitude, and travel times of teleseismic body waves. *Bulletin of the Seismological Society of America*, **103**(1), 268–282.

Garcia, R., and Souriau, A. 2000. Inner core anisotropy and heterogeneity level. *Geophysical Research Letters*, **27**(19), 3121–3124.

Geballe, Z. M., Lasbleis, M., Cormier, V. F., and Day, E. A. 2013. Sharp hemisphere boundaries in a translating inner core. *Geophysical Research Letters*, **40**, 1719–1723.

Giardini, D., Li, X.-D., and Woodhouse, J. H. 1987. Three-dimensional structure of the Earth from splitting in free-oscillation spectra. *Nature*, **325**, 405–411.

Giardini, D., Li, X.-D., and Woodhouse, J. H. 1988. Splitting functions of long-period normal modes of the Earth. *Journal of Geophysical Research*, **93**, 13716–13742.

Glatzmaier, G. A., and Roberts, P. H. 1996. Rotation and magnetism of Earth's inner core. *Science*, **274**(5294), 1887–1891.

Green, P. G. 1995. Reversible jump Markov chain Monte Carlo computation and Bayesian model determination. *Biometrika*, **82**, 711–732.

Gubbins, D. 1981. Rotation of the inner core. *Journal of Geophysical Research: Solid Earth*, **86**(B12), 11695–11699.

Gubbins, D., Masters, G., and Nimmo, F. 2008. A thermochemical boundary layer at the base of Earth's outer core and independent estimate of core heat flux. *Geophysical Journal International*, **174**(3), 1007–1018.

Gubbins, D., Sreenivasan, B., Mound, J., and Rost, S. 2011. Melting of the Earth's inner core. *Nature*, **473**(7347), 361–363.

Gutenberg, B. 1914. Über Erdbenwellen VIIA. Beobachtungen an Registrierungen von Fernbeben in Göttingen und Folgerungen über die Konstitution der Erdkörpers. *Nachrichten von der Gesellschaft der Wissenschaften zu Göttingen, Mathematisch-Physikalische Klasse*, 125–176.

Halley, E. 1686. An account of the cause of the change of the variation of the magnetical needle; with an hypothesis of the structure of the internal parts of the Earth: as it was proposed to the Royal Society in one of their late meetings. *Philosophical Transactions*, **16**(179–191), 563–578.

He, X., and Tromp, J. 1996. Normal-mode constraints on the structure of the Earth. *Journal of Geophysical Research: Solid Earth*, **101**(B9), 20053–20082.

Helffrich, G., and Sacks, S. 1994. Scatter and bias in differential PKP travel times and implications for mantle and core phenomena. *Geophysical Research Letters*, **21**(19), 2167–2170.

Helffrich, G., Kaneshima, S., and Kendall, J.-M. 2002. A local, crossing-path study of attenuation and anisotropy of the inner core. *Geophysical Research Letters*, **29**(12), 9-1–9-4.

Holme, R., and de Viron, O. 2005. Geomagnetic jerks and a high-resolution length-of-day profile for core studies. *Geophysical Journal International*, **160**, 435–439.

Huang, H.-H., Lin, F.-C., Tsai, V. C., and Koper, K. D. 2015. High-resolution probing of inner core structure with seismic interferometry. *Geophysical Research Letters*, **42**, 10622–10630.

Iritani, R., Takeuchi, N., and Kawakatsu, H. 2010. Seismic attenuation structure of the top half of the inner core beneath the northeastern Pacific. *Geophysical Research Letters*, **37**(19), L19303.

Iritani, R., Takeuchi, N., and Kawakatsu, H. 2014. Intricate heterogeneous structures of the top 300 km of the Earth's inner core inferred from global array data: I. Regional 1D attenuation and velocity profiles. *Physics of the Earth and Planetary Interiors*, **230**(0), 15–27.

Irving, J. C. E., and Deuss, A. 2011. Hemispherical structure in inner core velocity anisotropy. *Journal of Geophysical Research: Solid Earth*, **116**(B4), B04307.

Irving, J. C. E., Deuss, A., and Woodhouse, J. H. 2009. Normal mode coupling due to hemispherical anisotropic structure in Earth's inner core. *Geophysical Journal International*, **178**(2), 962–975.

Ishii, M., and Dziewoński, A. M. 2002. The innermost inner core of the Earth: Evidence for a change in anisotropic behavior at the radius of about 300 km. *PNAS*, **99**(22), 14026–14030.

Ishii, M., and Dziewoński, A. M. 2003. Distinct seismic anisotropy at the centre of the Earth. *Physics of the Earth and Planetary Interiors*, **140**, 203–217.

Ishii, M., and Dziewoński, A. M. 2005. Constraints on the outer-core tangent cylinder using normal-mode splitting measurements. *Geophysical Journal International*, **162**(3), 787–792.

Ishii, M., Tromp, J., Dziewoński, A. M., and Ekström, G. 2002a. Joint inversion of normal mode and body wave data for inner core anisotropy 1. Laterally homogeneous anisotropy. *Journal of Geophysical Research: Solid Earth*, **107**(B12), ESE 20–1–ESE 20–16.

Ishii, M., Dziewoński, A. M., Tromp, J., and Ekström, G. 2002b. Joint inversion of normal mode and body wave data for inner core anisotropy 2. Possible complexities. *Journal of Geophysical Research: Solid Earth*, **107**(B12), ESE 21–1–ESE 21–17.

Isse, T., and Nakanishi, I. 2002. Inner-core anisotropy beneath Australia and differential rotation. *Geophysical Journal International*, **151**, 255–263.

Jeanloz, R., and Wenk, H.-R. 1988. Convection and anisotropy of the inner core. *Geophysical Research Letters*, **15**(January), 72–75.

Jeffreys, H. 1926a. On the amplitudes of bodily seismic waves. *Geophysical Journal International*, **1**, 334–348.

Jeffreys, H. 1926b. The rigidity of the Earth's central core. *Geophysical Supplements to the Monthly Notices of the Royal Astronomical Society*, **1**(7), 371–383.

Jeffreys, H. 1939. The times of the core waves. *Monthly Notices of the Royal Astronomical Society Geophysical Supplement*, **4**, 548.

Jeffreys, H. 1976. *The Earth: Its Origin, History and Physical Constitution*, 6th edn. Cambridge University Press.

Jiang, G., and Zhao, D. 2012. Observation of high-frequency PKiKP in Japan: Insight into fine structure of inner core boundary. *Journal of Asian Earth Sciences*, **59**(0), 167–184. Minerals, Rocks and Mountains: Linking Petrology, Geochemistry and Geochronology.

Julian, B. R., Davies, D., and Sheppart, R. M. 1972. PKJKP. *Nature*, **235**(5337), 317–318.

Kaneshima, S. 1996. Mapping heterogeneity of the uppermost inner core using two pairs of core phases. *Geophysical Research Letters*, **23**(22), 3075–3078.

Kárason, H., and van der Hilst, R. D. 2001. Tomographic imaging of the lowermost mantle with differential times of refracted and diffracted core phases (PKP, Pdiff). *Journal of Geophysical Research: Solid Earth*, **106**(B4), 6569–6587.

Karato, S. 1993. Importance of anelasticity in the interpretation of seismic tomography. *Geophysical Research Letters*, **20**, 1623–1626.

Karato, S. 1999. Seismic anisotropy of the Earth's inner core resulting from flow induced by Maxwell stresses. *Nature*, **402**(December), 871–873.

Karato, S. 2003. Inner core anisotropy due to magnetic field-induced preferred orientation of iron. *Science*, **262**, 1708–1711.

Kawakatsu, H. 2006. Sharp and seismically transparent inner core boundary region revealed by an entire network observation of near-vertical PKiKP. *Earth Planets Space*, **58**(July), 855–863.

Kennett, B. L. N. 1998. On the density distribution within the Earth. *Geophysical Journal International*, **132**(2), 374–382.

Kennett, B. L. N. 2001. *The Seismic Wavefield: Volume 1, Introduction and Theoretical Development*, 1st edn. Cambridge University Press.

Kennett, B. L. N. 2002. *The Seismic Wavefield: Volume 2, Interpretation of Seismograms on Regional and Global Scales*, illustrated edn. Cambridge University Press.

Kennett, B. L. N. 2009. *Seismic Wave Propagation in Stratified Media*. Canberra, Australia: ANU Press.

Kennett, B. L. N., Engdahl, E. R., and Buland, R. 1995. Constraints on seismic velocities in the Earth from traveltimes. *Geophysical Journal International*, **122**, 108–124.

Kennett, B. L. N., Stipčević, J., and Gorbatov, A. 2015. Spiral-arm seismic arrays. *Bulletin of the Seismological Society of America*, **105**(4), 2109–2116.

Komatitsch, D., Ritsema, J., and Tromp, J. 2002. The spectral-element method, Beowulf computing, and three-dimensional seismology. *Science*, **298**, 1737–1742.

Koper, K. D., and Dombrovskaya, M. 2005. Seismic properties of the inner core boundary from PKiKP/P amplitude ratios. *Earth and Planetary Science Letters*, **237**(3–4), 680–694.

Koper, K. D., and Leyton, F. 2006. Decorrelation of coda waves from earthquake doublets recorded at YKA: Inner core super-rotation? In: *18th Annual Workshop*. Incorporated Research Institutions for Seismology, Tucson, Arizona.

Koper, K. D., and Pyle, M. L. 2004. Observations of PKiKP/PcP amplitude ratios and implications for Earth structure at the boundaries of the liquid core. *Journal of Geophysical Research: Solid Earth*, **109**.

Krasnoshchekov, D. N., Kaazik, P. B., and Ovtchinnikov, V. M. 2005. Seismological evidence for mosaic structure of the surface of the Earth's inner core. *Nature*, **435**(May), 483–487.

Kuang, W., and Bloxham, J. 1997. An Earth-like numerical dynamo model. *Nature*, **389**(6649), 371–374.

Labrosse, S., Poirier, J.-P., and Le Mouël, J.-L. 2001. The age of the inner core. *Earth and Planetary Science Letters*, **190**(3–4), 111–123.

Laske, G., and Masters, G. 1999. Limits on differential rotation of the inner core from an analysis of the Earth's free oscillations. *Nature*, **402**(6757), 66–69.

Laske, G., and Masters, G. 2003. *Earth's Core: Dynamics, Structure, Rotation*, 1st edn. American Geophysical Union Geodynamics Series, vol. 31. Washington, DC: American Geophysical Union Books Board. Chap. The Earth's free oscillations and the differential rotation of the inner core, pages 5–22.

Lay, T., and Wallace, T. C. 1995. *Modern Global Seismology*, 1st edn. San Diego: Academic Press.

Lee, T. D. 1957. *Nobel Lectures in Physics (1942–1962)*. 5 Toh Tuck Link, Singapore: World Scientific Publishing, Published for the Nobel foundation in 1998. Chap. Weak interactions and nonconservation of parity, Nobel Lecture, page 417.

Lehmann, I. 1936. P'. *Publications du Bureau Central Séismologique International*, **A14(3)**, S.87–115.

Lehmann, I. 1987. Seismology in the days of old. *Eos (Transactions, American Geophysical Union)*, **68**(3), 33–35.

Leykam, D., Tkalčić, H., and Reading, A. M. 2010. Core structure re-examined using new teleseismic data recorded in Antarctica: Evidence for, at most, weak cylindrical seismic anisotropy in the inner core. *Geophysical Journal International*, **180**(3), 1329–1343.

Leyton, F., and Koper, K. D. 2007a. Using PKiKP coda to determine inner core structure: 1. Synthesis of coda envelopes using single-scattering theories. *Journal of Geophysical Research (Solid Earth)*, **112**(May), 5316.

Leyton, F., and Koper, K. D. 2007b. Using PKiKP coda to determine inner core structure: 2. Determination of Q_C. *Journal of Geophysical Research (Solid Earth)*, **112**(May), 5317.

Leyton, F., Koper, K. D., Zhu, L., and Dombrovskaya, M. 2005. On the lack of seismic discontinuities within the inner core. *Geophysical Journal International*, **162**(3), 779–786.

Li, A., and Richards, P. G. 2003. Using earthquake doublets to study inner core rotation and seismicity catalog precision. *Geochemistry, Geophysics, Geosystems*, **4**(9), 1–23.

Li, D., Sun, D., and Helmberger, D. 2014. Notes on the variability of reflected inner core phases. *Earthquake Science*, **27(4)**, 441–468.

Li, X., and Cormier, V. F. 2002. Frequency-dependent seismic attenuation in the inner core, 1. A viscoelastic interpretation. *Journal of Geophysical Research (Solid Earth)*, **107**(December), 2361.

Li, X.-D., Giardini, D., and Woodhouses, J. H. 1991. Large-scale three-dimensional even-degree structure of the Earth from splitting of long-period normal modes. *Journal of Geophysical Research: Solid Earth*, **96**(B1), 551–577.

Lin, F.-C., and Tsai, V. C. 2013. Seismic Interferometry with antipodal station pairs. *Geophysical Research Letters*, **40**, 4609–4613.

Lin, F.-C., Tsai, V. C., Schmandt, B., and Duputel, Z. 2013. Extracting seismic core phases with array interferometry. *Geophysical Research Letters*, **40**, 1049–1053.

Lincot, A., Merkel, S., and Cardin, P. 2015. Is inner core seismic anisotropy a marker for plastic flow of cubic iron? *Geophysical Research Letters*, **42**, 1326–1333.

Lincot, A., Cardin, Ph., Deguen, R., and Merkel, S. 2016. Multiscale model of global inner-core anisotropy induced by hcp alloy plasticity. *Geophysical Research Letters*, **43**(3), 1084–1091. 2015GL067019.

Lindner, D., Song, X., Ma, P., and Christensen, D. H. 2010. Inner core rotation and its variability from nonparametric modeling. *Journal of Geophysical Research: Solid Earth*, **115**(B4).

Livermore, P. W., Hollerbach, R., and Jackson, A. 2013. Electromagnetically driven westward drift and inner-core superrotation in Earth's core. *Proceedings of the National Academy of Sciences*, **110**(40), 15914–15918.

Loper, D. E. 1983. Structure of the inner core boundary. *Geophysical & Astrophysical Fluid Dynamics*, **25**(1–2), 139–155.

Love, A. E. H. 1927. *A Treatise on the Mathematical Theory of Elasticity*, 4th edn. Cambridge University Press.

Lowrie, W. 2007. *Fundamentals of Geophysics*, 2nd edn. Cambridge University Press.

Lythgoe, K. H., Deuss, A., Rudge, J. F., and Neufeld, J. A. 2014. Earth's inner core: Innermost inner core or hemispherical variations? *Earth and Planetary Science Letters*, **385**, 181–189.

MacDonald, G. J. F., and Ness, N. F. 1961. A study of the free oscillations of the Earth. *Journal of Geophysical Research*, **66**(6), 1865–1911.

Mäkinen, A. M., and Deuss, A. 2011. Global seismic body-wave observations of temporal variations in the Earth's inner core, and implications for its differential rotation. *Geophysical Journal International*, **187**(1), 355–370.

Malinverno, A., and Briggs, V. 2004. Many potential solutions are generated with a variable number of unknowns using the reversible jump Monte Carlo algorithm. *Geophysics*, **69**, 1005.

Mao, H.-K., Shu, J., Shen, G., Hemley, R. J., Li, B., and Singh, A. K. 1998. Elasticity and rheology of iron above 220 GPa and the nature of the Earth's inner core. *Nature*, **396**(6713), 741–743.

Massé, R. P., Flinn, E. A., Seggelke, R. M., and Engdahl, E. R. 1974. PKIIKP and the average velocity of the inner core. *Geophysical Research Letters*, **1**(1), 39–42.

Masters, G., and Gilbert, F. 1981. Structure of the inner core inferred from observations of its spheroidal shear modes. *Geophysical Research Letters*, **8**(6), 569–571.

Masters, G., Jordan, T. H., Silver, P. G., and Gilbert, F. 1982. Aspherical Earth structure from fundamental spheroidal-mode data. *Nature*, **298**(5875), 609–613.

Masters, G., Laske, G., and Gilbert, F. 2000a. Autoregressive estimation of the splitting matrix of free-oscillation multiplet. *Geophysical Journal International*, **141**, 25–42.

Masters, G., Laske, G., and Gilbert, F. 2000b. Matrix autoregressive analysis of free-oscillation coupling and splitting. *Geophysical Journal International*, **143**, 478–489.

Masters, T. G. 1979. Observational constraints on the chemical and thermal structure of the Earth's deep interior. *Geophysical Journal International*, **57**, 507–534.

Masters, T. G., and Gubbins, D. 2003. On the resolution of density within the Earth. *Physics of the Earth and Planetary Interiors*, **140**, 159–167.

Mattesini, M., Belonoshko, A. B., Buforn, E., Ramírez, M., Simak, S. I., Udías, A., Mao, H.-K., and Ahuja, R. 2010. Hemispherical anisotropic patterns of the Earth's inner core. *Proceedings of the National Academy of Sciences*, **107**(21), 9507–9512.

Mattesini, M., Belonoshko, A. B., Tkalčić, H., Buforn, E., Udías, A., and Ahuja, R. 2013. Candy wrapper for the Earth's inner core. *Scientific Reports*, **3**, 2096.

McSweeney, T. J., Creager, K. C., and Merrill, R. T. 1997. Depth extent of inner-core seismic anisotropy and implications for geomagnetism. *Physics of the Earth and Planetary Interiors*, **101**(1–2), 131–156.

Merkel, S., Shu, J., Gillet, P., Mao, H.-K., and Hemley, R. J. 2005. X-ray diffraction study of the single-crystal elastic moduli of ϵ-Fe up to 30 GPa. *Journal of Geophysical Research: Solid Earth*, **110**(B5).

Mizzon, H., and Monnereau, M. 2013. Implication of the lopsided growth for the viscosity of Earth's inner core. *Earth and Planetary Science Letters*, **361**(0), 391–401.

Mohorovičić, A. 1910. Potres of 8. X 1909. *Godišnje izvješće Zagrebačkog meteorološkog opservatorija za godinu 1909*, **9**(4), 1–56.

Mohorovičić, A. 1913. Development in seismology in the last fifty years (Razvoj seizmologije posljednjih pedeset godina). *Ljetopis JAZU (in Croatian)*, **27**.

Mohorovičić, S. 1927. Über Nahbeben und über die Konstitution des Erdund Mondinnern. *Gerlands Beiträge zur Geophisik*, **XVII. Band**(1), 180–231.

Monnereau, M., Calvet, M., Margerin, L., and Souriau, A. 2010. Lopsided Growth of Earth's Inner Core. *Science*, **328**(5981), 1014–1017.

Morelli, A., and Dziewoński, A. M. 1987. Topography of the core-mantle boundary and lateral heterogeneity of the inner core. *Nature*, **325**, 678–683.

Morelli, A., Dziewonski, A. M., and Woodhouse, J. H. 1986. Anisotropy of the inner core inferred from PKIKP travel times. *Geophysical Research Letters*, **13**(13), 1545–1548.

Newton, I. 1687. *Philosophiae Naturalis Principia Mathematica*. J. Societatis Regiae ac Typis J. Streater.

Niazi, M., and Johnson, L. R. 1992. Q in the inner core. *Physics of the Earth and Planetary Interiors*, 55–62.

Nimmo, F., Price, G. D., Brodholt, J., and Gubbins, D. 2004. The influence of potassium on core and geodynamo evolution. *Geophysical Journal International*, **156**(2), 363–376.

Nishida, K. 2013. Global propagation of body waves revealed by cross-correlation analysis of seismic hum. *Geophysical Research Letters*, **40**, 1691–1696.

Niu, F., and Chen, Q.-F. 2008. Seismic evidence for distinct anisotropy in the innermost inner core. *Nature Geoscience*, **1**(10), 692–696.

Niu, F., and Wen, L. 2001. Hemispherical variations in seismic velocity at the top of the Earth's inner core. *Nature*, **410**(April), 1081–1084.

Niu, F., and Wen, L. 2002. Seismic anisotropy in the top 400 km of the inner core beneath the 'eastern' hemisphere. *Geophysical Research Letters*, **29**(June), 1611.

Ohtaki, T., Kaneshima, S., and Kanjo, K. 2012. Seismic structure near the inner core boundary in the south polar region. *Journal of Geophysical Research: Solid Earth*, **117**(B3).

Okal, E. A., and Cansi, Y. 1998. Detection of PKJKP at intermediate periods by progressive multi-channel correlation. *Earth and Planetary Science Letters*, **164**(1–2), 23–30.

Oldham, R. D. 1900. On the propagation of earthquake motion to great distances. *Philosophical Transactions of the Royal Society of London A: Mathematical, Physical and Engineering Sciences*, **194**(252–261), 135–174.

Oldham, R. D. 1906. The constitution of the interior of the Earth, as revealed by earthquakes. *Quarterly Journal of the Geological Society*, **62**(1–4), 456–475.

Ouzounis, A., and Creager, K. C. 2001. Isotropy overlying anisotropy at the top of the inner core. *Geophysical Research Letters*, **28**(22), 4331–4334.

Pachhai, S., Tkalčić, H., and Masters, G. 2016. Estimation of splitting functions from Earth's normal mode spectra using neighbourhood algorithm. *Geophysical Journal International*, **204**(1).

Peng, Z., Koper, K. D., Vidale, J. E., Leyton, F., and Shearer, P. 2008. Inner-core fine-scale structure from scattered waves recorded by LASA. *Journal of Geophysical Research (Solid Earth)*, **113**(September), 9312.

Poli, P., Campillo, M., Pedersen, H., and Group, LAPNET Working. 2012. Body-wave imaging of Earth's mantle discontinuities from ambient seismic noise. *Science*, **338**, 1063–1065.

Poupinet, G, and Kennett, B. L. N. 2004. On the observation of high frequency PKiKP and its coda in Australia. *Physics of the Earth and Planetary Interiors*, **146**(3–4), 497–511.

Poupinet, G., Pillet, R., and Souriau, A. 1983. Possible heterogeneity of the Earth's core deduced from PKIKP travel times. *Nature*, **305**, 204–206.

Poupinet, G., Souriau, A., and Coutant, O. 2000. The existence of an inner core super-rotation questioned by teleseismic doublets. *Physics of the Earth and Planetary Interiors*, **118**(1–2), 77–88.

Poynting, J. H. 1891. On a determination of the mean density of the Earth and the gravitation constant by means of the common balance. *Philosophical Transactions of the Royal Society of London. (A.)*, **182**, 565–656.

Press, W. H. 2007. *Numerical Recipes, 3rd edition: The Art of Scientific Computing*, 3rd edn. Cambridge University Press.

Qamar, A. 1973. Revised velocities in the Earth's core. *Bulletin of the Seismological Society of America*, **63**(3), 1073–1105.

Resovsky, J. S., and Ritzwoller, M. H. 1998. New and refined constraints on three-dimensional Earth structure from normal modes below 3 mHz. *Journal of Geophysical Research: Solid Earth*, **103**(B1), 783–810.

Rial, J. A., and Cormier, V. F. 1980. Seismic waves at the epicenter's antipode. *Journal of Geophysical Research*, **85**, 2661–2668.

Richards, P. G. 1976. On the adequacy of plane-wave reflection/transmission coefficients in the analysis of seismic body waves. *Bulletin of the Seismological Society of America*, **66**(3), 701–717.

Richards, P. G., Song, X., and Li, A. 1998. Detecting possible rotation of Earth's inner core. *Science*, **282**(November), 1227a.

Ritzwoller, M., Masters, G., and Gilbert, F. 1986. Observations of anomalous splitting and their interpretation in terms of aspherical structure. *Journal of Geophysical Research: Solid Earth*, **91**(B10), 10203–10228.

Ritzwoller, M., Masters, G., and Gilbert, F. 1988. Constraining aspherical structure with low-degree interaction coefficients: Application to uncoupled multiplets. *Journal of Geophysical Research: Solid Earth*, **93**(B6), 6369–6396.

Romanowicz, B., Tkalčić, H., and Bréger, L. 2003. *Earth's Core: Dynamics, Structure, Rotation*, 1st edn. American Geophysical Union Geodynamics Series, vol. 31. Washington, DC: American Geophysical Union Books Board. Chap. on the origin of complexity in PKP travel time data, pages 31–44.

Romanowicz, B., and Bréger, L. 2000. Anomalous splitting of free oscillations: A reevaluation of possible interpretations. *Journal of Geophysical Research: Solid Earth*, **105**(B9), 21559–21578.

Romanowicz, B., and Mitchell, B. J. 2015. *Treatise on Geophysics*, 2nd edn. Vol. 1: Deep Earth Seismology. Amsterdam: Elsevier B.V. Chap. Deep earth structure: Q of the Earth from crust to core, pages 789–828.

Romanowicz, B., Li, X.-D., and Durek, J. 1996. Anisotropy in the inner core: Could it be due to low-order convection? *Science*, **274**, 963–966.

Romanowicz, B., Cao, A., Godwal, B., Wenk, R., Ventosa, S., and Jeanloz, R. 2016. Seismic anisotropu in the Earth's innermost inner core: Testing structural models against mineral physics predictions. *Geophysical Research Letters*, **43**(1), 93–100.

Rost, S., and Thomas, C. 2002. Array seismology: methods and applications. *Reviews of Geophysics*, **40**(3), 1–27.

Ruigrok, E., Draganov, D., and Wapenaar, K. 2008. Global-scale seismic interferometry: Theory and numerical examples. *Geophysical Prospecting*, **56**, 395–417.

Sambridge, M. 1999. Geophysical inversion with a neighbourhood algorithm I. Searching a parameter space. *Geophysical Journal International*, **103**, 4839–4878.

Sambridge, M., Gallagher, K., Jackson, A., and Rickwood, A. 2006. Trans-dimensional inverse problems, model comparison and the evidence. *Geophysical Journal International*, **167**, 528–542.

Sambridge, M., Bodin, T., Gallagher, K., and Tkalčić, H. 2013. Transdimensional inference in the geosciences. *Philosophical Transactions of the Royal Society A: Mathematical, Physical and Engineering Sciences*, **371**(1984), 20110547.

Schuster, G. T. 2009. *Seismic Interferometry*. Cambridge University Press.

Sha, X., and Cohen, R. E. 2010. Elastic isotropy of ϵ-Fe under Earth's core conditions. *Geophysical Research Letters*, **37**(10).

Shapiro, N. M., Campillo, M., Stehly, L., and Ritzwoller, M. H. 2005. High-resolution surface-wave tomography from ambient seismic noise. *Science*, **307**, 1615–1618.

Sharrock, D. S., and Woodhouse, J. H. 1998. Investigation of time dependent inner core structure by the analysis of free oscillation spectra. *Earth Planets Space*, **50**, 1013–1018.

Shearer, P., and Masters, G. 1990. The density and shear velocity contrast at the inner core boundary. *Geophysical Journal International*, **102**(2), 491–498.

Shearer, P. M. 1994. Constraints on inner core anisotropy from PKP(DF) travel times. *Journal of Geophysical Research*, **99**(B10), 19647–19659.

Shearer, P. M. 2009. *Introduction to Seismology*, 2nd edn. Cambridge University Press.

Shearer, P. M., and Toy, K. M. 1991. PKP(BC) versus PKP(DF) differential travel times and aspherical structure in the Earth's inner core. *Journal of Geophysical Research*, **96**(B2), 2233–2247.

Shearer, P. M., Toy, K. M., and Orcutt, J. A. 1988. Axi-symmetric Earth models and inner-core anisotropy. *Nature*, **333**(May), 228–232.

Shearer, P. M., Rychert, C. A., and Liu, Q. 2011. On the visibility of the inner-core shear wave phase PKJKP at long periods. *Geophysical Journal International*, **185**(3), 1379–1383.

Singh, S. C., Taylor, M. A. J., and Montagner, J. P. 2000. On the presence of liquid in Earth's inner core. *Science*, **287**(5462), 2471–2474.

Snieder, R. 2004. Extracting the Green's function from the correlation of coda waves: A derivation based on stationary phase. *Physical Review E*, **69**.

Snieder, R., Grêt, A., Douma, H., and Scales, J. 2002. Coda wave intereferometry for estimating nonlinear behaviour in seismic velocity. *Science*, **295**, 2253–2255.

Song, X. 1996. Anisotropy in central part of inner core. *Journal of Geophysical Research*, **101**(B7), 16089–16097.

Song, X. 2000a. Joint inversion for inner core rotation, inner core anisotropy, and mantle heterogeneity. *Journal of Geophysical Research*, **105**(B4), 7931–7943.

Song, X. 2000b. Time dependence of PKP(BC)-PKP(DF) times: Could it be an artifact of potential systematic earthquake mislocations? *Physics of the Earth and Planetary Interiors*, **122**, 221–228.

Song, X., and Helmberger, D. V. 1993. Anisotropy of Earth's inner core. *Geophysical Research Letters*, **20**(23), 2591–2594.

Song, X., and Helmberger, D. V. 1995. Depth dependence of anisotropy of Earth's inner core. *Journal of Geophysical Research*, **100**(B7), 9805–9816.

Song, X., and Helmberger, D. V. 1998. Seismic evidence for an inner core transition zone. *Science*, **282**, 924–927.

Song, X., and Li, A. 2000. Support for differential inner core superrotation from earthquakes in Alaska recorded at South Pole station. *Journal of Geophysical Research: Solid Earth*, **105**(B1), 623–630.

Song, X., and Poupinet, G. 2007. Inner core rotation from event-pair analysis. *Earth and Planetary Science Letters*, **261**, 259–266.

Song, X., and Richards, P. G. 1996. Seismological evidence for differential rotation of the Earth's inner core. *Nature*, **382**, 221–224.

Song, X., and Xu, X. 2002. Inner core transition zone and anomalous PKP(DF) waveforms from polar paths. *Geophysical Research Letters*, **29**(4), 1–1–1–4.

Souriau, A. 1989. A search for time dependent phenomena inside the core from seismic data. In: *EGS Meeting Abstracts*.

Souriau, A. 1998a. Detecting possible rotation of Earth's inner core: Response. *Science*, **282**, 1227a.

Souriau, A. 1998b. Is the rotation real? *Science*, **281**, 55–56.

Souriau, A. 1998c. New seismological constraints on differential rotation of the inner core from Novaya Zemlya events recorded at DRV, Antarctica. *Geophysical Journal International*, **134**(2), F1–F5.

Souriau, A., and Calvet, M. 2015. *Treatise on Geophysics*, 2nd edn. Vol. 1: Deep Earth Seismology. Amsterdam: Elsevier B.V. Chap. Deep earth structure: The Earth's cores, pages 725–757.

Souriau, A., and Poupinet, G. 2000. Inner core rotation: A test at the worldwide scale. *Physics of the Earth and Planetary Interiors*, **118**, 13–27.

Souriau, A., and Romanowicz, B. 1996. Anistropy in inner core attenuation: A new type of data to constrain the nature of the solid core. *Geophysical Research Letters*, **23**(1), 1–4.

Souriau, A., and Romanowicz, B. 1997. Anisotropy in the inner core: Relation between P-velocity and attenuation. *Physics of the Earth and Planetary Interiors*, **101**, 33–47.

Souriau, A., and Roudil, P. 1995. Attenuation in the uppermost inner core from broad-band GEOSCOPE PKP data. *Geophysical Journal International*, **123**, 572–587.

Souriau, A., and Souriau, M. 1989. Ellipticity and density at the inner core boundary from subcritical PKiKP and PcP data. *Geophysical Journal International*, **98**(1), 39–54.

Souriau, A., Roudil, P., and Moynot, B. 1997. Inner core differential rotation: Facts and artefacts. *Geophysical Research Letters*, **24**(16), 2103–2106.

Souriau, A., Teste, A., and Chevrot, S. 2003. Is there any structure inside the liquid outer core? *Geophysical Research Letters*, **30**(11).

Stein, S., and Wysession, M. 2003. *An Introduction to Seismology, Earthquakes, and Earth Structure*. Oxford, UK: Blackwell Publishing Ltd.

Steinle-Neumann, G., Stixrude, L., Cohen, R. E., and Gulseren, O. 2001. Elasticity of iron at the temperature of the Earth's inner core. *Nature*, **413**(6851), 57–60.

Stevenson, D. J. 1981. Models of the Earth's core. *Science*, **214**(4521), 611–619.

Stevenson, D. J. 1987. Limits on lateral density and velocity variations in the Earth's outer core. *Geophysical Journal of the Royal Astronomical Society*, **88**(1), 311–319.

Stipčević, J., Tkalčić, H., Herak, M., Markušić, S., and Herak, D. 2011. Crustal and uppermost mantle structure beneath the external Dinarides, Croatia, determined from teleseismic receiver functions. *Geophysical Journal International*, **185**, 1003–1019.

Stixrude, L., and Cohen, R. E. 1995. High-pressure elasticity of iron and anisotropy of Earth's inner core. *Science*, **267**(5206), 1972–1975.

Stroujkova, A., and Cormier, V. F. 2004. Regional variations in the uppermost 100 km of the Earth's inner core. *Journal of Geophysical Research (Solid Earth)*, **109**(October), 10307.

Su, W.-J., and Dziewonski, A. M. 1995. Inner core anisotropy in three dimensions. *Journal of Geophysical Research*, **100**(B7), 9831–9852.

Su, W.-J., Dziewonski, A. M., and Jeanloz, R. 1996. Planet within a planet: Rotation of the inner core of Earth. *Science*, **274**(5294), 1883–1887.

Sumita, I., and Olson, P. 1999. A laboratory model for convection in Earth's core driven by a thermally heterogeneous mantle. *Science*, **286**(5444), 1547–1549.

Sumita, I., Yoshida, S., Kumazawa, M., and Hamano, Y. 1996. A model for sedimentary compaction of a viscous medium and its application to inner-core growth. *Geophysical Journal International*, **124**(2), 502–524.

Sun, X., and Song, X. 2008. Tomographic inversion for three-dimensional anisotropy of Earth's inner core. *Physics of the Earth and Planetary Interiors*, **167**(1–2), 53–70.

Sun, X., Poupinet, G., and Song, X. 2006. Examination of systematic mislocation of South Sandwich Islands earthquakes using station pairs: Implications of inner core rotation. *Journal of Geophysical Research*, **111**(B11305).

Sylvander, M., and Souriau, A. 1996. P-velocity structure of the core-mantle boundary region inferred from PKP(AB)-PKP(BC) differential travel times. *Geophysical Research Letters*, **23**, 853–856.

Tanaka, S. 2012. Depth extent of hemispherical inner core from PKP(DF) and PKP(Cdiff) for equatorial paths. *Physics of the Earth and Planetary Interiors*, **210–211**(0), 50–62.

Tanaka, S., and Hamaguchi, H. 1997. Degree one heterogeneity and hemispherical variation of anisotropy in the inner core from PKP(BC)-PKP(DF) times. *Journal of Geophysical Research*, **102**(February), 2925–2938.

Tanaka, S., and Tkalčić, H. 2015. Complex inner core boundary from frequency characteristics of the reflection coefficients of PKiKP waves observed by Hi-net. *Progress in Earth and Planetary Science*, **2**, 1–16.

Tateno, S., Hirose, K., Ohishi, Y., and Tatsumi, Y. 2010. The structure of iron in Earth's inner core. *Science*, **330**(6002), 359–361.

Tkalčić, H. 2015. Complex inner core of the Earth: The last frontier of global seismology. *Reviews of Geophysics*, **53**(1), 59–94. 2014RG000469.

Tkalčić, H., Flanagan, M. P., and Cormier, V. F. 2006. Observation of near-podal P'P' precursors: Evidence for back scattering from the 150–220 km zone in the Earth's upper mantle. *Geophysical Research Letters*, **33**(3).

Tkalčić, H., Kennett, B. L. N., and Cormier, V. F. 2009. On the inner–outer core density contrast from PKiKP/PcP amplitude ratios and uncertainties caused by seismic noise. *Geophysical Journal International*, **179**(1), 425–443.

Tkalčić, H., Young, M., Muir, J. B., Davies, D. R., and Mattesini, M. 2015. Strong, Multi-Scale Heterogeneity in Earth's Lowermost Mantle. *Scientific Reports*, **5**(12), 18416 EP –.

Tkalčić, H., and Flanagan, M. P. 2004 (December). Structure of the deep inner core from antipodal PKPPKP waves. Pages Abstract T54A–06 of: *Eos Trans. AGU, Fall Meet. Suppl.*, vol. 85. American Geophysical Union.

Tkalčić, H. 2001 (December). *Study of deep Earth structure using body waves.* PhD thesis, University of California at Berkeley.

Tkalčić, Hrvoje. 2010. Large variations in travel times of mantle-sensitive seismic waves from the South Sandwich Islands: Is the Earth's inner core a conglomerate of anisotropic domains? *Geophysical Research Letters*, **37**.

Tkalčić, H., and Kennett, B. L. N. 2008. Core structure and heterogeneity: A seismological perspective. *Australian Journal of Earth Sciences*, **55**(4), 419–431.

Tkalčić, H., Romanowicz, B., and Houy, N. 2002. Constraints on D" structure using PKP(AB-DF), PKP(BC-DF) and PcP-P traveltime data from broad-band records. *Geophysical Journal International*, **148**, 599–616.

Tkalčić, H., Cormier, V., Kennett, B. L. N., and He, K. 2010. Steep reflections from the Earth's core reveal small-scale heterogeneity in the upper mantle. *Physics of the Earth and Planetary Interiors*, **178**, 80–91.

Tkalčić, H., Bodin, T., Young, M., and Sambridge, M. 2013a. On the nature of the P-wave velocity gradient in the inner core beneath Central America. *Journal of Earth Science*, **24**(5), 699–705.

Tkalčić, H., Young, M., Bodin, T., Ngo, S., and Sambridge, M. 2013b. The shuffling rotation of the Earth's inner core revealed by earthquake doublets. *Nature Geoscience*, **6**(6), 497–502.

Torsvik, T. H., Smethurst, M. A., Burke, K., and Steinberger, B. 2006. Large igneous provinces generated from the margins of the large low-velocity provinces in the deep mantle. *Geophysical Journal International*, **167**(3), 1447–1460.

Tromp, J. 1993. Support for anisotropy of the Earth's inner core from free oscillations. *Nature*, **366**, 679–681.

Tromp, J. 1995. Normal-mode splitting observations from the Great 1994 Bolivia and Kuril earthquakes: Constraints on the structure of the mantle and inner core. *GSA Today*, **5**, 137–151.

Tromp, J., and Zanzerkia, E. 1995. Toroidal splitting observations from the Great 1994 Bolivia and Kuril Islands earthquakes. *Geophysical Research Letters*, **22**(16), 2297–2300.

Verhoogen, J. 1961. Heat balance of the earth's core. *Geophysical Journal of the Royal Astronomical Society*, **4**, 276–281.

Verne, J. G. 1864. *Voyage au centre de la Terre (Journey to the Centre of the Earth).*

Vidale, J. E., and Earle, P. S. 2000. Fine-scale heterogeneity in the Earth's inner core. *Nature*, **404**(March), 273–275.

Vidale, J. E., and Earle, P. S. 2005. Evidence for inner-core rotation from possible changes with time in PKP coda. *Geophysical Research Letters*, **32**(1). L01309.

Vidale, J. E., Dodge, D. A., and Earle, P. S. 2000. Slow differential rotation of the Earth's inner core indicated by temporal changes in scattering. *Nature*, **405**(6785), 445–448.

Vinnik, L., Romanowicz, B., and Breger, L. 1994. Anisotropy in the center of the inner core. *Geophysical Research Letters*, **21**(16), 1671–1674.

Vočadlo, L., Alfe, D., Gillan, M. J., Wood, I. G., Brodholt, J. P., and Price, G. D. 2003. Possible thermal and chemical stabilization of body-centred-cubic iron in the Earth's core. *Nature*, **424**(6948), 536–539.

Vočadlo, L., Dobson, D. P., and Wood, I. G. 2009. Ab initio calculations of the elasticity of hcp-Fe as a function of temperature at inner-core pressure. *Earth and Planetary Science Letters*, **288**(3–4), 534–538.

Wang, T., Song, X., and Xia, H. H. 2015. Equatorial anisotropy in the inner part of Earth's inner core from autocorrelation of earthquake coda. *Nature Geoscience*, **8**(3), 224–227.

Waszek, L., and Deuss, A. 2011. Distinct layering in the hemispherical seismic velocity structure of Earth's upper inner core. *Journal of Geophysical Research*, **116**, B12313.

Waszek, L., and Deuss, A. 2015. Observations of exotic inner core waves. *Geophysical Journal International*, **200**, 1636–1650.

Waszek, L., Irving, J., and Deuss, A. 2011. Reconciling the hemispherical structure of Earth's inner core with its super-rotation. *Nature Geoscience*, **4**(February), 264–267.

Wedemeyer-Bohm, S., Scullion, E., Steiner, O., van der Voort, L. R., de la Cruz Rodriguez, J., Fedun, V., and Erdelyi, R. 2012. Magnetic tornadoes as energy channels into the solar corona. *Nature*, **486**(7404), 505–508.

Wen, L. 2006. Localized temporal change of the Earth's inner core boundary. *Science*, **314**, 967–970.

Wen, L., and Niu, F. 2002. Seismic velocity and attenuation structures in the top of the Earth's inner core. *Journal of Geophysical Research (Solid Earth)*, **107**(November), 2273.

Whaler, K., and Holme, R. 1996. Catching the inner core in a spin. *Nature*, **382**, 205–206.

Widmer, R., Masters, G., and Gilbert, F. 1991. Spherically symmetric attenuation within the Earth from normal mode data. *Geophysical Journal International*, **104**(3), 541–553.

Widmer, R., Masters, G., and Gilbert, F. 1992. Observably split multiplets – data analysis and interpretation in terms of large-scale aspherical structure. *Geophysical Journal International*, **111**(3), 559–576.

Wiechert, E. 1897. Ueber die massenverteilung im inneren der erde. *Nachrichten von der Gesellschaft der Wissenschaften zu Göttingen, Mathematisch-Physikalische Klasse*, **1897**, 221–243.

Woodhouse, J. H. 1988. *The calculation of eigenfrequencies and eigenfunctions of the free oscillations of the Earth and the Sun*. Vol. Seismological algorithms, Computational methods and computer programs. London, UK: Academic Press.

Woodhouse, J. H., and Dahlen, F. A. 1978. The effect of a general aspherical perturbation on the free oscillations of the Earth. *Geophysical Journal of the Royal Astronomical Society*, **53**, 335–354.

Woodhouse, J. H., and Girnius, T. 1982. Surface waves and free oscillations in a regionalized Earth model. *Geophysical Journal of the Royal Astronomical Society*, **68**, 653–673.

Woodhouse, J. H., Giardini, D., and Li, X.-D. 1986. Evidence for inner core anisotropy from free oscillations. *Geophysical Research Letters*, **13**(13), 1549–1552.

Wookey, J., and Helffrich, G. 2008. Inner-core shear-wave anisotropy and texture from an observation of PKJKP waves. *Nature*, **454**(7206), 873–876.

Xu, X., and Song, X. 2003. Evidence for inner core super-rotation from time-dependent differential PKP traveltimes observed at Beijing Seismic Network. *Geophysical Journal International*, **152**(3), 509–514.

Yee, T.-G., Rhie, J., and Tkalčić, H. 2014. Regionally heterogeneous uppermost inner core observed with Hi-net array. *Journal of Geophysical Research: Solid Earth*, **119**(10), 7823–7845.

Yoshida, S., Sumita, I., and Kumazawa, M. 1996. Growth model of the inner core coupled with the outer core dynamics and the resulting elastic anisotropy. *Journal of Geophysical Research*, **101**(December), 28085–28104.

Yu, W.-C., Wen, L., and Niu, F. 2005. Seismic velocity structure in the Earth's outer core. *Journal of Geophysical Research (Solid Earth)*, **110**(February), 2302.

Zeng, X., and Ni, S. 2013. Constraining shear wave velocity and density contrast at the inner core boundary with PKiKP/P amplitude ratio, *Journal of Earth Science*, **24**(5), 716–724.

Zhang, J., Song, X., Li, Y., Richards, P. G., Sun, X., and Waldhauser, F. 2005. Inner core differential motion confirmed by earthquake waveform doublets. *Science*, **309**(5739), 1357–1360.

Zhang, J., Richards, P. G., and Schaff, D. P. 2008. Wide-scale detection of earthquake waveform doublets and further evidence for inner core super-rotation. *Geophysical Journal International*, **174**(3), 993–1006.

Zou, Z., Koper, K. D., and Cormier, V. F. 2008. The structure of the base of the outer core inferred from seismic waves diffracted around the inner core. *Journal of Geophysical Research: Solid Earth*, **113**, B05314.

Index

Printed in the United States
by Baker & Taylor Publisher Services